PRAISE FOR *GALILEO*

One of *Symmetry* magazine's Top Physics Books of the Year

"One would have hoped that the Galileo story could be treated just as the fascinating history this book makes clear it is—but we really *need* this story now, because we're living through the next chapter of science denial, with stakes that couldn't be higher."

—Bill McKibben, author of *Falter: Has the Human Game Begun to Play Itself Out?*

"Every so often a reason arises to retell the life of Galileo. This year, as Mario Livio so forcefully demonstrates in *Galileo and the Science Deniers,* the 400-year Galileo Affair casts an urgent new light on the current climate crisis."

—Dava Sobel, author of *Longitude, Galileo's Daughter,* and *The Glass Universe*

"Few historical episodes are more fraught with subtleties, ironies and ambiguities. To tell it properly requires an unusual breadth of knowledge and the gifts of a great storyteller. Fortunately, Mario Livio is fully equipped for the task. . . . [Livio] tells the story of Galileo in a perceptive, illuminating and balanced way."

—Stephen M. Barr, *The Washington Post*

"Livio illuminates the parallels between the deniers of Galileo's scientific findings and those today who ignore the evidence of climate change. Intriguing and accessible, and packed with clever insights, Livio's latest gives readers plenty to think about."

—*Publishers Weekly*

"This is an insightful, riveting, and deeply researched biography of Galileo Galilei that reveals not just his complex character but also how he was truly intellectually radical and well ahead of his time."

—Priyamvada Natarajan, professor, Departments of Astronomy and Physics, Yale University, and author of *Mapping the Heavens: The Radical Scientific Ideas That Reveal the Cosmos*

GALILEO
AND THE
SCIENCE
DENIERS

MARIO LIVIO

Simon & Schuster Paperbacks

NEW YORK   LONDON   TORONTO
SYDNEY   NEW DELHI

Simon & Schuster Paperbacks
An Imprint of Simon & Schuster, Inc.
1230 Avenue of the Americas
New York, NY 10020

First Simon & Schuster trade paperback edition May 2021

SIMON & SCHUSTER PAPERBACKS and colophon are registered
trademarks of Simon & Schuster, Inc.

For information about special discounts for bulk purchases, please contact
Simon & Schuster Special Sales at 1-866-506-1949
or business@simonandschuster.com.

The Simon & Schuster Speakers Bureau can bring authors to your
live event. For more information or to book an event, contact the
Simon & Schuster Speakers Bureau at 1-866-248-3049 or
visit our website at www.simonspeakers.com.

Text design by Paul Dippolito

Manufactured in the United States of America

1 3 5 7 9 10 8 6 4 2

The Library of Congress has cataloged the hardcover edition as follows:

Names: Livio, Mario, 1945– author.
Title: Galileo and the denialists / Mario Livio.
Description: First Simon & Schuster hardcover edition. | New York : Simon
& Schuster, 2020. | Includes bibliographical references and index.
Identifiers: LCCN 2019032142 | ISBN 9781501194733 (hardcover) | ISBN 9781501194788 (ebook)
Subjects: LCSH: Galilei, Galileo, 1564–1642. | Astronomers—Italy—Biography. |
Scientists—Italy—Biography. | Religion and science. | Faith and reason—Christianity.
Classification: LCC QB36.G2 L658 2020 | DDC 520.92 [B]—dc23
LC record available at https://lccn.loc.gov/2019032142

ISBN 978-1-5011-9473-3
ISBN 978-1-5011-9474-0 (pbk)
ISBN 978-1-5011-9478-8 (ebook)

To Sofie

Contents

CHAPTER 15   The Final Years   207

CHAPTER 16   The Saga of Pio Paschini   213

CHAPTER 17   Galileo's and Einstein's Thoughts
             on Science and Religion   221

CHAPTER 18   One Culture   233

             Acknowledgments   239

             Notes   241

             Bibliography   257

             Photo Credits   271

             Index   273

Preface

Being an astrophysicist myself, I have always been fascinated by Galileo. He was, after all, not only the founder of modern astronomy and astrophysics—the person who turned an ancient profession into a window onto the universe's deepest secrets and awe-inspiring wonders—but also a symbol of the fight for intellectual freedom.

Using a simple arrangement of lenses fixed at the two ends of a hollow cylinder, Galileo was able to revolutionize our understanding of the cosmos and of our place within it. Fast-forward four centuries, and we find a great-great-great-grandson of Galileo's telescope: the Hubble Space Telescope.

Over the decades during which I worked as a scientist with Hubble (till 2015), I was often asked what I thought gave the Hubble telescope its iconic status as one of the most recognizable projects in scientific history. I've identified at least six main reasons for Hubble's popularity. In no particular order, these are:

- The incredible images produced by the space telescope, dubbed by one journalist "the Sistine Chapel of the scientific age."

- The actual scientific discoveries to which Hubble has significantly contributed. Those range from determining the composition of the atmospheres of extrasolar planets to the astounding discovery that the cosmic expansion is accelerating.

- The drama associated with the telescope. The transformation of what was initially considered a disastrous failure—a flaw in the telescope's mirror was discovered within weeks after its launch—into a gigantic success.

- The ingenuity of scientists and engineers, coupled with the courage of astronauts, all of which helped to overcome the incredible technological challenges involved in making repairs and upgrades several hundred miles above the Earth.

- The telescope's longevity: it was launched in 1990 and is still working beautifully in 2019.

- An extraordinarily effective dissemination and outreach program, which circulates the findings to scientists, to the general public, and to educators, in an efficient, attractive, and easily accessible fashion.

Amazingly, when I carefully examined Galileo's life and work, I realized that the same key words came to mind: *images, discoveries, drama, ingenuity, courage, longevity,* and *dissemination.*

First, Galileo created breathtaking images from his observations of the lunar surface. Second, while his spectacular discoveries about the solar system and the Milky Way didn't conclusively prove that the world was Copernican, with the Earth revolving around the Sun, they all but destroyed the stability of the Earth-centered Ptolemaic universe.

Finally, the drama characterizing Galileo's life, the brilliant ingenuity he showed in his experiments in mechanics, the courage he demonstrated in defending his views, his enormous success in disseminating his findings and in making them accessible, and the fact that his ideas became the basis on which modern science has been erected, are the main characteristics that make Galileo and his story immortal.

You may wonder why I felt absolutely compelled to write another book about Galileo, when quite a few excellent biographies and analyses of his work exist already. There were three main reasons for my decision. First, I realized that very few of the known biographies were written by a research astronomer or astrophysicist. I believe, or at least hope, that someone actively engaged in astrophysical research can bring a novel perspective and fresh insights even to this seemingly overworked arena. In particular, I have attempted in this book to place Galileo's discoveries in the context of today's knowledge, ideas, and intellectual setting.

Second, and most important, I am convinced that present-day readers will be amazed to discover how relevant Galileo's story is for today. In a world of governmental antiscience attitudes with science deniers at key positions, unnecessary conflicts between science and religion, and the perception of a widening schism between the humanities and the sciences, Galileo's tale serves, first of all, as a potent reminder of the importance of freedom of thought. At the same time, Galileo's complex personality itself, grounded as it was in late-Renaissance Florence, Italy, provides a perfect example of the fact that all the achievements of the human mind are part of just *one* culture.

Finally, many of the superb, scholarly written biographies include parts that are rather abstruse or far too detailed even for educated but nonexpert readers. My goal has been to provide an accurate yet relatively short and accessible account of the life and work of this captivating man. In some sense, I am humbly attempting to follow here in Galileo's footsteps. He insisted on publishing many of his scientific findings in Italian (rather than Latin), for the benefit of every educated person rather than for a limited elite. I hope to do the same for Galileo's tale and its vitally important message.

CHAPTER I

Rebel with a Cause

At a breakfast that took place at the Medici Palace in Pisa, Italy, in December 1613, Galileo's former student Benedetto Castelli was asked to explain the significance of Galileo's discoveries with the telescope. During the discussion that ensued, the Grand Duchess Christina of Lorraine badgered Castelli about what she perceived as contradictions between certain biblical passages and the Copernican view of an Earth orbiting a stationary Sun. She cited in particular the description from the book of Joshua, in which, at Joshua's request, the Lord commanded the Sun (and not the Earth) to stand still over the ancient Canaanite city of Gibeon and the Moon to stop in its course over the Aijalon Valley. Castelli described the entire affair in a letter he sent to Galileo on December 14, 1613, claiming that he played the theologian "with such assurance and dignity" that it would have done Galileo good to hear him. Overall, Castelli summarized, he "carried things off like a paladin."

Galileo was apparently less convinced of his student's success in elucidating the issue, since in a long letter to Castelli that he sent on December 21, he explained in detail his own views on the impropriety of using Scripture to dispute science: "I believe that the authority of Holy Writ had only the aim of persuading men of those articles and propositions which, being necessary for our salvation and overriding all human reason, could not be made credible by any other science," he wrote. In a style characterizing much of his writing, he was quick to add sarcastically that he did not think "that the same God who has given us our senses, reason, and intelligence wished us to abandon their use." Simply put, Galileo argued that when an apparent conflict arises between Scripture and what experience and demonstration establish about nature, Scripture has to be reinterpreted in an alternative way. "Especially," he noted, "in matters of which only a minimal part, and in partial conclusion, is to be read in Scripture, for such is astronomy, of which there is [in the Bible] so small a part that not even the planets are named."

While the argument itself was not entirely new—theologian Saint Augustine had written already in the fifth century that the sacred writers did not intend to teach science, "since such knowledge was of no use to salvation"—Galileo's bold statements were about to put him on a collision course with the Catholic Church. The *Letter to Benedetto Castelli* marked only the beginning of the risky road that would eventually lead to Galileo being pronounced "vehemently suspected of heresy" on June 22, 1633. Overall, if we examine the record of Galileo's life in terms of his personal contentment, it traces something like an inverted-U shape, with a pronounced peak somewhere shortly after his numerous astronomical discoveries, followed by a fairly steep fall. Ironically, the parabolic paths of projectiles, which Galileo was the first to determine, form a similar curve.

As history would have it, Galileo's tragic end only helped to transform him into one of those larger-than-life heroes of our intellectual history. There aren't many scientists, after all, about whose lives and achievements entire plays (such as Bertolt Brecht's unforgettable *Life of Galileo*, first performed in 1943), and scores of poems have been written, or an opera has been composed. Suffice it also to note that a Google search on "Galileo Galilei" produced no fewer than 36 million results, again demonstrating an impact that many of today's academics would love to have.

Albert Einstein once wrote about Galileo that "he is the father of modern physics—indeed, of modern science altogether." He was echoing here philosopher and mathematician Bertrand Russell, who also called Galileo "the greatest of the founders of modern science." Einstein added that Galileo's "discovery and use of scientific reasoning" was "one of the most important achievements in the history of human thought." These two thinkers were not in the habit of offering profuse praise, but there was a solid base for these accolades. Through his pioneering, stubborn insistence that the book of nature was "written in the language of mathematics," and his successful fusion of experimentation, idealization, and quantification, Galileo literally reshaped natural history. He transformed it from being a mere collection of vague, verbal, nebulous accounts embellished by metaphors, to a magnificent opus encompassing (when the contemporary knowledge allowed it) rigorous mathematical theories. Within those theories, observations, experiments, and reasoning became the only acceptable methods for discovering facts about the world and for investigating new connections in nature. As Max Born, winner of the 1954 Nobel Prize in Physics, once put it: "The scientific attitude and methods of experimental and theoretical research have been the same through the centuries since Galileo and will remain so."

His scientific prowess notwithstanding, we should not get the impression that Galileo was the easiest or kindest person, or, for that matter, even that he was an idealistic freethinker; an explorer who accidentally wandered into theological controversy. Whereas he could indeed be extremely empathic and supportive to members of his own family, he showed blistering intolerance and belligerence, wielding his sharp pen toward scientists who disagreed with him. A number of scholars labeled Galileo a zealot, although not always a zealot for the same cause. Some said it was for Copernicanism—the scheme in which the Earth and the other planets revolve around the Sun—others claimed he was a zealot for his own self-righteousness. Still others even believed he was fighting for the Catholic Church, anxious to stop it from making a mistake of historical proportions by condemning a scientific theory that he was convinced would be proven to represent a correct description of the cosmos. In defense of his zeal, though, one would probably expect nothing less from a man who set out not only to change a worldview that had existed for centuries but also to introduce entirely new approaches to what constitutes scientific knowledge.

Undoubtedly, Galileo owes much of his scholarly fame to his spectacular discoveries with the telescope and his extremely effective dissemination of his findings. Turning this new device to the heavens instead of watching sailing ships or his neighbors, he was able to show wonders such as: there are mountains on the surface of the Moon; Jupiter has four satellites orbiting it; Venus displays a series of changing phases like the Moon; and the Milky Way is composed of a vast number of stars. But even these proverbially out-of-this-world achievements are not sufficient to explain the enormous popularity that Galileo enjoys to this very day, and the fact that he, more than almost any other scientist (with the possible exceptions of Sir Isaac Newton and Einstein), has become the perennial symbol of scientific imagination and courage. In addition, the facts that Galileo was the first to firmly establish the laws of falling bodies and the founder of the crucial concept of dynamics in physics were clearly not enough to make him the hero of the scientific revolution. What at the end distinguished Galileo from most of his contemporaries was not so much what he believed in but rather why he believed it and how he reached that belief.

Galileo based his convictions on experimental evidence (sometimes real, sometimes in the form of "thought experiments"—thinking through the consequences of a hypothesis) and theoretical contemplation, and not on authority. He was prepared to recognize and internalize that what had been trusted for centuries might be wrong. He also had the foresight to assert forcefully that the road to scientific truth is paved with patient experimentation leading to mathematical laws that weave all the observed facts into one harmonious tapestry. As such, he can definitely be regarded as one of the inventors of what we call today the scientific method: a sequence of steps that ideally (although rarely in reality) needs to be taken for the development of a new theory, or for acquiring more advanced knowledge. The Scottish empiricist philosopher David Hume gave in 1759 this personal comparison between Galileo and another famous empiricist, English philosopher and statesman Francis Bacon: "Bacon pointed out at a distance the road to true philosophy: Galileo both pointed it out to others, and made himself considerable advances in it. The Englishman was ignorant of geometry; the Florentine revived that science, excelled in it, and was the first to apply it, together with experiment, to natural philosophy."

All of Galileo's impressive insights could not have happened in a vacuum. One could perhaps even argue that the age shapes individuals more than individuals shape the age. Art historian Heinrich Wolfflin wrote once: "Even the most original talent cannot proceed beyond certain limits which are fixed for it by the date of its birth." What, then, was the backdrop against which Galileo acted and produced his unique magic?

Galileo was born in 1564, only a few days before the death of the great artist Michelangelo (and also the same year that brought the world the playwright William Shakespeare). He died in 1642, almost one year before the birth of Newton. One doesn't have to believe in the transmigration at death of the soul of one human into a new body—nobody should—to realize that the torch of culture, knowledge, and creativity is always passed from one generation to the next.

Galileo was, in many respects, an example of a product of the late Renaissance. In the words of Galileo scholar Giorgio de Santillana: "a classic type of humanist, trying to bring his culture to the awareness of the new scientific ideas." Galileo's last disciple and first biographer (or perhaps more of a hagiographer), Vincenzo Viviani, wrote about his master: "he praised the good things that had been written in philosophy and in geometry to elucidate and awaken the mind to their own order of thinking and maybe higher, *but* he said that the main entrance to the very rich treasure of material philosophy was *observations and experiments*, which through the senses as keys, could reach the most noble and inquisitive intellects." Precisely the same sentiments had been expressed by the great polymath Leonardo da Vinci about a century earlier, when he defied those who had mocked him for not being "well read," by exclaiming: "Those who study the ancients and not the works of Nature are stepsons and not sons of Nature, the mother of all good authors." Viviani further tells us that the judgment Galileo passed on various works of art was highly valued by celebrated artists such as the painter and architect Lodovico Cigoli, who was Galileo's personal friend and sometimes collaborator. Indeed, apparently in response to a request from Cigoli, Galileo wrote an essay in which he discussed the superiority of painting over sculpture. Even the famous Baroque painter Artemisia Gentileschi approached Galileo when she thought that the French noble Charles de Lorraine, 4th Duke of Guise, had not sufficiently appreciated one of her paintings. Moreover, in her painting *Judith Slaying Holofernes*, her depiction of blood squirting was in accordance with Galileo's discovery of the parabolic trajectory of projectiles.

Viviani's encomium doesn't stop there. His plaudits just go on and on. In a style very reminiscent of that of the first art historian, Giorgio Vasari, in his biographies of the greatest painters, Viviani writes that Galileo was a superb lutenist whose playing "surpassed in beauty and grace even that of his father." This particular praise appears to have been at least somewhat misplaced: while it is true that Galileo's father, Vincenzo Galilei, was a composer, lutenist, and music theorist, and that Galileo himself played the lute quite well, it was Galileo's younger brother Michelangelo who was a true lute virtuoso.

Finally, to top it all, Viviani relates that Galileo could recite at length by heart from the works of the famous Italian poets Dante Alighieri, Ludovico Ariosto, and Torquato Tasso. This was not exaggerated adulation. Galileo's favorite poem truly was Ariosto's *Orlando Furioso*, a rich, chivalric fantasy, and he devoted a serious literary work to a comparison between Ariosto and Tasso, in which he extolled

Ariosto while brutally criticizing Tasso. He once told his neighbor (and later biographer) Niccolò Gherardini that reading Tasso after Ariosto was like eating sour lemons after delicious melons. True to his Renaissance spirit, Galileo continued to be deeply interested in art and in contemporary poetry throughout his entire life, and his writings, even on scientific matters, both reflected and were informed by his literary erudition.

In addition to this splendid artistic and humanistic background, there were, of course, important scientific advances—a few genuinely revolutionary—that helped pave the way for the type of conceptual breakthroughs that Galileo was about to produce. The year 1543, in particular, witnessed the publication of not one but two books that were about to change humanity's views on both the microcosm and the macrocosm. Nicolaus Copernicus published *On the Revolutions of the Heavenly Spheres*, which proposed to demote the Earth from its central position in the solar system, and the Flemish anatomist Andreas Vesalius published *On the Fabric of the Human Body*, in which he presented a new understanding of human anatomy. Both books went against prevailing beliefs that had dominated thought since antiquity. Copernicus's book inspired others, such as philosopher Giordano Bruno and later astronomers Johannes Kepler and indeed Galileo himself, to expand the Copernican heliocentric ideas even further. Similarly, by elbowing out ancient authorities such as the Greek physician Galen, Vesalius's book incentivized William Harvey, the first anatomist to recognize the full circulation of blood in the human body, to advocate the primacy of visual evidence. Major advances happened in other branches of science as well. The English physicist William Gilbert published his influential book on the magnet in 1600, and the Swiss physician Paracelsus introduced in the sixteenth century a new perspective on diseases and toxicology.

All of these discoveries created a certain openness to science not seen in the earlier Dark Ages. Still, the intellectual outlook of even the most educated people at the end of the sixteenth century was predominantly medieval. This was about to change dramatically in the seventeenth century. There must have been additional factors, therefore, that were responsible for what we might call the "Galileo phenomenon." Other things ought to have been radically revised to create the fertile ground that was eventually ready to receive Galileo and promote him to the status of protomartyr and an icon of scientific freedom.

An important new sociopsychological element in the late sixteenth and early seventeenth centuries was the rise of *individualism*—the notion that a person can achieve self-fulfillment irrespective of social circumstances. This novel perspective manifested itself in areas ranging from the acquisition of knowledge to the accumulation of wealth, and from the determination of moral truths to the evaluation of entrepreneurial success. The individualist attitude was very different from the values inherited from the ancient Greek philosophy, in which people were considered primarily members of the larger community rather than individuals. Plato's *The Republic*, for instance, aimed to define and help construct a superior society, not a better person.

During the Middle Ages, individualism was prevented from taking root by the actions of the Catholic Church, through the principle that truths and ethics were determined by religious councils composed of a collection of "wise men" rather than by the experiences, contemplations, or opinions of freethinkers. This type of dogmatic rigidity started to crack with the rise of the Protestant movements, which rebelled against the assertion that those councils were infallible. Ideas espoused by the ensuing Reformation war penetrated other areas of culture. The war was waged not only on the battlefield and with propaganda pamphlets, one-page broadsheets, and essays, but also with paintings by artists such as Lucas Cranach the Elder, that contrasted Protestant and Catholic Christianity. It was partly the diffusion of these individualist convictions into philosophy that enabled the Galileo phenomenon. The same ideas were later put squarely center stage by the French philosopher René Descartes, who argued that an individual's thoughts are the best, if not the only, proof of existence. ("I think, therefore I am.")

There was also a new technology—printing—that made possible both the individual's access to knowledge and the standardization of information. The invention of movable type and the printing press in mid-fifteenth-century Europe had an immense impact. Literacy was suddenly not the preserve of a rich elite, and the dissemination of data and scholarship through printed books continuously increased the numbers of educated people. But that was not all. More people, from different walks of life, were now exposed to precisely the *same* books, leading to the establishment of a new information basis and a more democratic education. In the seventeenth century, students of botany, astronomy, anatomy, or even the Bible in, say, Rome could be using the same texts as their counterparts in Venice or Prague.

The resemblance of this proliferation of sources of information to the effects and ramifications of the internet, social media, and communication devices today immediately jumps to mind. As an early precursor to e-mail, Twitter, Instagram, and Facebook, printing also allowed individuals to transmit their ideas to the masses more rapidly and efficiently. When the German theologian Martin Luther campaigned for church reform, he was assisted greatly by the existence of printing. In particular, his translation of the Bible from Latin into German vernacular, to represent his ideal of a world in which ordinary people could consult the word of God for themselves, had a profound impact on both the modern German language and the Church in general. About two hundred thousand copies in hundreds of reprinted editions appeared before Luther's death. Similarly, no scientist at the time had a greater talent than Galileo for communicating his discoveries to others. Convinced that his message was ushering in a new science, he saw his role as that of the great persuader, and printing books in Italian rather than in the traditional Latin (which benefited only a few learned individuals) proved to be a potent tool to this end.

Perhaps less obvious was the fact that printing also had an effect on mathematics. The ability to relatively easily reproduce diagrams, coupled with the printing of classical Greek manuscripts, renewed interest in Euclidean geometry, which Galileo was to make creative use of. Archimedes, the greatest mathematician of antiquity, would become his role model. Among many other achievements, Archimedes formulated the law of the lever and used it capably against the Romans in his legendary war machines. "Give me a place to stand, and I will move the Earth!" he was reported to have exclaimed. Galileo was only too happy to demonstrate that most machines could, at their basic principles, be reduced to something resembling a lever. Eventually he also came to believe in the Copernican model, in which the Earth was moving even without human intervention.

More broadly, the recovery, fresh editing, and translation of texts from the classical past provided a basis for more skeptical, investigative, observational attitudes. The primacy of mathematics as key to both practical and theoretical advances was becoming apparent, and it burgeoned into Galileo's guiding light. Mathematics proved essential in areas ranging from painting (where it was used for working out vanishing points and foreshortening in perspective) to business transactions (where mathematician Luca Pacioli introduced double-entry accounting in his influential book *The Collected Knowledge of Arithmetic, Geometry, Proportion and Proportionality*). The upsurge in the numerical thinking of the time was perhaps best illustrated by an amusing anecdote involving Lord Burghley (William Cecil), the chief advisor to Queen Elizabeth I of England. According to this story, in 1555 he took the surprising step of weighing himself, his wife, his son, and all his household servants, and listing all the results.

Finally, another factor that helped to enhance the reverberations of Galileo's findings was the intense curiosity about newly discovered worlds brought about by the great explorers. Together with the geographical horizons, the span of knowledge also rolled wider starting with the last decade of the fifteenth century. Explorers such as Christopher Columbus, John Cabot, and Vasco da Gama reached the Caribbean islands, landed in North America, and found the sea route to India, respectively, just between 1492 and 1498. Then, by the 1520s, humans had circled the globe. No wonder that when the nineteenth-century French historian Jules Michelet tried to summarize the thirst for new wisdom and humanism that characterized the Renaissance, he concluded that it encompassed "the discovery of the world and of man."

A MAN OF HIS TIME AND BEFORE HIS TIME

Galileo's journey as a scientist started in 1583, when he dropped out of medical school and began to study mathematics. By 1590, at the age of twenty-six he already had the audacity to criticize the teachings on motion of the great Greek philosopher Aristotle, according to which things moved because of a built-in impetus. About thirteen years later, following a series of ingenious experiments with inclined planes and pendulums, Galileo formulated the very first "laws of motion" concerning free fall, even though he would not publish those until 1638.

He presented his first breathtaking discoveries with the telescope in 1610, and five years later, in a famous *Letter to the Grand Duchess Christina*, expressed his risky opinion that the biblical language had to be interpreted in light of what science reveals, and not the other way around.

In spite of his personal disagreements with some orthodox church dicta, as late as May 18, 1630, Galileo was still received in Rome as an honored guest by Pope Urban VIII, and he left the city under the impression that the Pope had approved the printing of his book *Dialogue Concerning the Two Chief World Systems* after only a few minor corrections and a change of title. Overestimating the strength of his friendship with the pontiff and underestimating the fragility of the delicate psychological and political position of the Pope in that turbulent post-Reformation era, Galileo continued to believe that reason would prevail. "Facts, which at first seem improbable, will, even on scant explanation, drop the cloak which has hidden them and stand forth in naked and simple beauty," he once wrote. Imprudently neglecting his own safety, he proceeded to get the book to print, and, after a rather convoluted series of events, the book finally went to press on February 21, 1632. Whereas in the preface to the book Galileo purported to discuss the Earth's motion merely as a "mathematical caprice," the text itself had a very different flavor. In fact, Galileo taunted and derided those who still refused to accept the Copernican view in which the Earth revolved around the Sun.

Einstein said about this book:

> [It] is a mine of information for anyone interested in the cultural history of the Western world and its influence upon economic and political development. A man is here revealed who possesses the passionate will, the intelligence, and the courage to stand up as the representative of rational thinking against the host of those who, relying on the ignorance of the people and the indolence of teachers in priest's and scholar's garb, maintain and defend their positions of authority.

For Galileo, however, the publication of the *Dialogo*, as it is commonly referred to, marked the beginning of the end of his life, though not of his fame. He was tried by the inquisition in 1633, pronounced a suspected heretic, forced to recant his Copernican ideas, and eventually placed under house arrest. The *Dialogo* was put on the Vatican's *Index of Prohibited Books*, where it remained until 1835.

In 1634 Galileo suffered another devastating blow with the death of his beloved daughter Sister Maria Celeste. He still managed to write one more book, *Discourses and Mathematical Demonstrations Concerning Two New Sciences* (commonly known as *Discorsi*), which was smuggled out of Italy to Holland and published there in Leiden. The book summarized much of his life's work, from his early days in Pisa, some fifty years earlier. Although his own travel was forbidden, Galileo was allowed to have occasional visitors. One of his callers during that late period of his life was the young John Milton, of *Paradise Lost* fame.

Galileo died in 1642 at his villa in Arcetri, near Florence, after having been blind and bedridden for a while. But as we shall clearly see in this book, his science and the tale of Galileo and his times resonate strongly today. There is a striking similarity between some of the religious, social, economic, and cultural problems that a person in the seventeenth century had to struggle with, and those we encounter in the twenty-first century. Indeed, whose story is better to tell than that of Galileo if we are to shine light on current concerns such as the continuing debate about the proper realms of science and religion, the support for the teaching of creationist ideas, and the uninformed attacks on intellectualism and expertise? The blatant dismissal in some circles of the research on climate change, the mocking attitude directed at the funding of basic research, and the elimination of budgets for the arts and public radio in the United States are only a few of the manifestations of such assaults.

There are additional reasons why Galileo and his seventeenth-century world are extremely relevant for us and our cultural needs. An important one is the apparent schism between the sciences and the humanities first identified and exposed in a 1959 talk (and later a book) by British physical chemist and novelist C. P. Snow, with his coinage of the term "the Two Cultures." Snow presented his concern with great clarity: "A good many times, I have been present at gatherings of people who, by the standards of the traditional culture, are thought highly educated and who have with considerable gusto been expressing their incredulity at the illiteracy of scientists." At the same time, Snow pointed out, had he asked those very same erudite essayists to define *mass* or *acceleration*—to him, the scientific equivalent of "Can you read?"—for nine in ten of the highly educated, he might as well have been speaking a foreign language. On the whole, Snow noted that during the 1930s and onward, literary scholars started referring to themselves as "the intellectuals," thereby excluding scientists from this coterie. Some of those intellectuals even resented the penetration of scientific methods into areas not traditionally associated with the exact sciences, such as sociology, linguistics, and the arts. While surely not as extreme, their stance was not entirely dissimilar from the indignation expressed by church officials who reacted against what they regarded as Galileo's unwelcome intrusion into theology.

A few scholars argue that the problem of the two cultures is less acute today than it was when Snow gave his lecture. Others, however, claim that a proper dialogue between the two cultures is still mostly absent. Historian of science David Wootton, for example, feels that the problem has even deepened. In his book *The Invention of Science: A New History of the Scientific Revolution*, Wootton writes: "History of science, far from serving as a bridge between the arts and sciences, nowadays offers the scientists a picture of themselves that most of them cannot recognize."

In 1991 author and literary agent John Brockman introduced the concept of a "third culture," in online conversation and later in a book with that title. According to Brockman, the third culture "consists of those scientists and other thinkers in the empirical world who, through their work and expository writing, are taking the place of the traditional intellectuals in rendering visible the deeper meaning of our lives, redefining who and what we are." As we shall see in this book, four hundred years ago, Galileo would have secured himself a place of honor in the third culture.

The border between art and science was largely blurred during the Renaissance, with artists such as Leonardo da Vinci, Piero della Francesca, Albrecht Dürer, and Filippo Brunelleschi having been involved in serious scientific research or in mathematics. Consequently, Galileo himself embodied an integration of the humanities and the sciences that can serve as a model to be examined, even if not easily emulated today. Consider, for instance, that at age twenty-four, he presented two lectures on the topic of "On the Shape, Location, and Size of Dante's *Inferno*," or the fact that even Galileo's science involved, to a great extent, the visual arts. For example, in his book *The Sidereal Messenger* (*Sidereus Nuncius*), a booklet of sixty pages that was rushed to print in 1610, he tells his scientific story of the

Moon through a series of wonderful wash drawings, probably relying on the lessons in art he had received from the painter Cigoli at the Accademia delle Arti del Disegno (Academy of the Arts of Drawing) in Florence.

Perhaps most important, Galileo was the pioneer and star of advancing the new art of experimental science. He realized that he could test or suggest theories by artificially manipulating various terrestrial phenomena. He was also the first scientist whose vision and scientific outlook incorporated both methods and results that were applicable to all branches of science.

Galileo made numerous discoveries, but, in four areas, he literally revolutionized the field: astronomy and astrophysics; the laws of motion and mechanics; the astonishing relationship between mathematics and physical reality (dubbed in 1960 by physicist Eugene Wigner "the unreasonable effectiveness of mathematics"); and experimental science. Largely through his unparalleled intuition and partly through his training in chiaroscuro—the art of representing three dimensions in two through a clever use of light and shadows—he was able to transform what would have otherwise been simple visual experiences into intellectual conclusions about the heavens.

Following Galileo's numerous observations and the confirmation of his findings by other astronomers, no one could cogently argue anymore that what one saw through the telescope must have been an optical illusion and not a faithful reproduction of reality. The only defense remaining to those obstinately refusing to accept the conclusions implied by the accumulating weight of empirical facts and scientific reasoning was to reject the interpretation of the results almost solely on the basis of religious or political ideology. If such a reaction sounds disturbingly similar to the present-day denial by some people of the reality of climate change, or of the theory of evolution by means of natural selection, it's because it is!

CHAPTER 2

A Humanist Scientist

Galileo Galilei was born in Pisa on February 15 or 16, 1564. His mother, Giulia Ammannati, was an educated, if prickly, difficult, and bitter woman from Pescia, whose family was involved in the wool and clothing business. His father, Vincenzo, was a Florentine musician and music theorist from a family with noble roots, but who had rather unimpressive financial means. Even then, musicians struggled to support themselves and their families on their music alone, so Vincenzo apparently also became a part-time cloth merchant. The couple married in 1563, and, following Galileo, had two more sons and three or, according to some, four daughters. Of these, only Galileo's younger brother, Michelangelo, and the two sisters Livia and Virginia played significant roles in Galileo's life.

Genetics are inescapable. In Galileo's case, he may have inherited at least some of his rebellious nature, self-righteousness, and distrust of authority from his father, and his selfishness, jealousy, and anxiety from his mother. Vincenzo Galilei objected vehemently to the musical theory promoted by his own teacher, Gioseffo Zarlino. A music theorist from the old school, Zarlino was a strong advocate of a tradition that dated all the way back to the ancient Pythagoreans, according to which all string-generated sounds that are pleasant to our ears (such as the octave or the fifth) are the result of plucking identical strings, with lengths that are in ratios of integer numbers such as 1:2, 2:3, 3:4, and so on. It was the uncompromising clinging to this scheme that produced the old joke that Renaissance musicians spent half their time tuning their instruments and the other half playing out of tune.

Vincenzo, on the other hand, maintained that adhering to this conservative numerology was arbitrary and that other criteria, equally valid if not better, could be adopted. In simple terms, Galileo's father argued that musical consonance is determined by the musician's ear rather than by his or her arithmetical capabilities. By insisting on freeing music from Pythagoras, Vincenzo opened the door for the modern "equally tempered system" popularized later by Johann Sebastian Bach. Through a series of experiments with strings of different materials and of varying tensions, he showed, for example, that strings of different tensions could produce the octave at a length ratio different from the canonical 2:1 (which was used when the tension was held constant). Almost prophetically—or rather, probably having an influence on his son—Vincenzo entitled one of his books on the subject *Dialogue on Ancient and Modern Music*, and another *Discourse Concerning the Work of Messer Gioseffo Zarlino of Chioggia.* Years later, two of Galileo's most important books would be entitled *Dialogue Concerning the Two Chief World Systems* and *Discourses and Mathematical Demonstrations Concerning Two New Sciences.* One sentence in particular in Vincenzo's fictional dialogue on music precisely captured the credo that Galileo was about to espouse later in life. The two interlocutors agree from the outset that they should invariably "set aside . . . not only authority, but also reasoning that seems plausible but is contradictory to the perception of truth."

The young Galileo probably helped his father in the experiments with the strings, and in the process, he might have started to realize the importance of the evidence-based approach to science. This could have been the first step in Galileo becoming a firm believer in the concept that in trying to find descriptions of natural phenomena, one needs, as he later expressed it: "to seek out and clarify the definition that best agrees with that which nature employs." Having to perform a series of experiments with weights hung on strings (to vary the tension), may have also planted in his mind the seeds of the idea of using pendulums to measure time.

Vincenzo was not only a talented lutenist, and his interests went beyond his particular objections to contrapuntal polyphony. Besides being an active member of the Florentine Camerata—a group of cultured Florentine intelligentsia interested in music and literature—he was educated in the classical languages and in mathematics. In short, not just in terms of the period of time during which he happened to live, Vincenzo was quite what we would call today a Renaissance man.

Having grown up in that milieu, Galileo was about to follow in his father's intellectual footsteps—although not in the direction of music, even though he often played second lute with Vincenzo. At the same time, having also witnessed his father's idealistic ambitions being frustrated by harsh reality, especially economically, may have instilled in Galileo a stubborn, tenacious will to succeed.

Galileo's relationship with his mother was rather more problematic. Even Galileo's brother Michelangelo described her as an absolutely "terrible" woman. Yet, in spite of numerous unpleasant incidents that included Giulia spying on Galileo and attempting to steal a few of his telescope lenses in order to give them to her son-in-law, he did his best in later years to attend to her ever-growing pecuniary needs.

Galileo's father returned from Pisa to Florence when Galileo was about ten. Lack of space in the home of a financially strapped family, in which the number of children was rapidly increasing, may have been one of the reasons for leaving Galileo in Pisa for a while, to live with his mother's relative Muzio Tedaldi. His primary education at that stage was in what we normally refer to today as the liberal arts: Latin, poetry, and music. Both Galileo's first biographer, Viviani, and Galileo's neighbor and second biographer, Niccolò Gherardini, tell us that Galileo rapidly surpassed the level at which his teacher was able to help him and that he continued his schooling through reading classical authors by himself.

At age eleven, he was sent to the monastery at Vallombrosa, in the serene atmosphere of which he studied logic, rhetoric, and grammar. He was also exposed to the visual arts by virtue of observing the work of artists in residence at the monastery. At that impressionable age, he must have been inspired by the abbot of Vallombrosa, who was apparently a polymath with knowledge in fields ranging from mathematics, to astrology, to theology, as well as in "all the other sound arts and sciences."

While there is no doubt that Galileo found the intellectual and spiritual ambience at the monastery appealing, we don't know with certainty whether he truly intended to become a novice of the order of Camaldolese monks. Be that as it may, however, Vincenzo certainly had different plans for Galileo. Partly wanting perhaps to revive his family's glorious past, which included a great-grandfather who had been a famous Florentine medical doctor, but at the same time striving to ensure Galileo's economic future, Vincenzo enrolled his son as a medical student at the University of Pisa in September 1580.

Unfortunately, medicine, which at the time was being taught based primarily on the teachings of the celebrated anatomist from ancient Greece Galen of Pergamum and which was filled with rigid rules and superstitions, bored Galileo. He did not feel that he should "give himself up . . . almost blindly" to the assertions and opinions of archaic writers. However, something good did come out of his first years at Pisa: he met the Tuscan court mathematician Ostilio Ricci. After listening to Ricci's lectures on Euclidean geometry, Galileo was bewitched. In fact, according to Viviani, even earlier, "the great talent and delight that he had . . . in painting, perspective, and music, hearing his father frequently say that such things had their origin in geometry, moved in him a desire to try it." Consequently, he started to devote all of his time to studying Euclid on his own, while totally neglecting medicine.

More than three centuries later, Einstein would be quoted as saying: "If Euclid failed to kindle your youthful enthusiasm, then you were not born to be a scientific thinker." Galileo passed this particular "test" with flying colors. Moreover, envisaging mathematics as his vocation, he introduced Ricci to his father in the summer of 1583, hoping that the mathematician would convince Vincenzo that this was the right choice. Ricci explained to Vincenzo that mathematics was the topic Galileo was truly passionate about, and expressed his willingness to be the young man's instructor. Vincenzo, who was a fairly good mathematician himself, did not object in principle, but he had the legitimate fatherly concern that Galileo would not find a job in mathematics. After all, he himself had already experienced what it meant to have the not particularly remunerative profession of a musician. Therefore, he insisted that Galileo complete his studies of medicine first, threatening that he would close his purse if Galileo refused. Fortunately for the history of science, the father and son eventually reached a compromise: Galileo could continue his studies of mathematics for one more year, with his father's support, after which he would take on the obligation of sustaining himself.

Ricci introduced Galileo to the works of Archimedes, whose genius in applying mathematics to physics and to real-life engineering problems was to motivate Galileo and permeate his entire scientific work. Ricci's own professor, mathematician Niccolò Tartaglia, was the scholar who published a few of Archimedes's works in Latin, and he had also produced an authoritative Italian translation of Euclid's masterwork *The Elements*. Not surprisingly, two of Galileo's very first treatises—one that addressed the problem of finding the center-of-mass of a system of weights, and the other on the conditions under which bodies float in water—were both on topics in which Archimedes had shown great interest. Galileo's second biographer, Gherardini, cited Galileo as saying: "One could travel securely without hindrance through heaven and earth, if one only did not lose sight of the teachings of Archimedes." The ironic net result of this entire sequence of events in the young man's life, however, was that Galileo—one of the greatest scientific minds in history—left the University of Pisa in 1585, after having dropped out of medicine and not having completed any degree.

Nevertheless, Galileo's studies with Ricci and the introduction to Archimedes were not in vain. They inculcated in him a strong belief that mathematics can provide the necessary decoding tools for deciphering nature's secrets. Through mathematics, he saw a way to translate phenomena into precise statements that could then be tested and proven unambiguously. This insight was truly remarkable. About 350 years later, Einstein would still wonder, "How is it possible that mathematics, a product of human thought that is independent of experience, fits so excellently the objects of physical reality?"

Viviani tells a fascinating story about Galileo's time as a student in Pisa: In 1583, the nineteen-year-old watched a lamp suspended on a long chain in the cathedral at Pisa swing from side to side. Galileo realized, through counting his heartbeats, that the time it took the lamp to oscillate a full amplitude was constant (strictly speaking, only as long as the range of the swing was not too large). From this simple observation, Viviani writes admiringly, Galileo went on, and "by very precise experiments, he verified the equality of its [the pendulum's] vibrations" [the constancy of the period of the swinging]. Viviani recounts further that Galileo used this constancy of the period of a pendulum to construct a medical apparatus to measure the pulse rate. This tale became so widely known in later years that in 1840, painter Luigi Sabatelli created a beautiful fresco depicting young Galileo observing the lamp (see the leftmost panel at the top, in Figure 1 of the color insert).

There is only one "small" problem with this captivating account. The lamp in question was installed in the cathedral at Pisa only in 1587, four years after Galileo had supposedly watched it swing. It's possible, of course, that Galileo observed a different lamp that had hung earlier in the same spot. However, since Galileo himself mentions the constancy of the pendulum's swing for the first time only in 1598, and there is no documented evidence for him having invented any instrument to measure the pulse rate, most historians of science suspect that Viviani's colorful description of Galileo's precociousness was the type of embellishment typical for biographies at the time.

In reality, the Venetian physician Santorio Santorio did publish in 1626 the details of his *pulsilogium*—a device that could accurately measure the pulse rate based on the constant period of the pendulum. Galileo, who was usually very aggressive about any attempt to deny him credit, never claimed priority. Nevertheless, the fact that Galileo may have experimented in his father's workshop with weights hung on strings (which effectively constitute pendulums) does leave open the possibility of a grain of truth in Viviani's account. Galileo definitely did start to use pendulums as time-measuring devices in 1602, and he even had the idea for a pendulum clock in 1637. Galileo's son, Vincenzo, started constructing a model based on his father's concept, but, unfortunately, died before finishing, in 1649. Such a working clock was eventually invented in 1656 by the Dutch scientist Christiaan Huygens.

Having left Pisa without a degree, Galileo had to find some means to make ends meet, so he started to teach mathematics privately, partly in Florence and partly in Siena. In 1586 he also published a small scientific tract entitled *La Bilancetta* (*The Little Balance*), which was not particularly original, except for introducing a more accurate way of weighing objects in the air and in water. This was especially useful for jewelers, whose common practice was to weigh precious metals this way.

At the end of 1586, Galileo started to compose a treatise on motion and free-falling bodies. Following the ancient example of Plato, Galileo wrote in dialogue form. This genre was extremely popular in sixteenth-century Italy as a vehicle for technical exposition, polemic, and miniature dramas of persuasion. This book was never finished, and it addressed mostly problems that seem rather trivial by today's standards. Yet it constituted an important step along Galileo's road to a new mechanics. The book did contain, in particular, two interesting points. First, at age twenty-two, Galileo already had the chutzpah to challenge the great Aristotle on topics related to motion, even though the necessary mathematical tools to treat such variables as velocity and acceleration did not exist yet. (Calculus, which allowed for the proper definitions of velocity and acceleration as *rates* of changes, was formulated by Newton and Gottfried Leibniz only in the mid-seventeenth century.)

The second interesting point was that Galileo did reach the tentative conclusion that irrespective of their weight, falling bodies made of the same material move with the same speed in a given medium. In later years, this was going to be part of one of his major discoveries in mechanics.

Given the drama associated with Galileo's name and his acceptance of Copernicanism, it is also intriguing to discover that in a separate manuscript, *Treatise on the Sphere, or Cosmography*—probably written in the late 1580s and most likely intended primarily for his private teachings—Galileo fully adopted the old Ptolemaic geocentric system in which the Sun, the Moon, and all the planets revolved around the Earth in circular orbits. This was about to change drastically in the years to come.

In an attempt to beef up his still unremarkable résumé, Galileo paid a visit in 1587 to the foremost mathematician of the Jesuit order in Rome: Christopher Clavius. Clavius, who became a full member of the order in 1575, had been teaching the mathematical subjects at Rome's prestigious Collegio Romano since 1564. In 1582 he was the senior mathematician on the commission that instituted the Gregorian calendar. Galileo set his eyes on one position in particular: a chair of mathematics had opened up at the University of Bologna, the oldest university in the Western world, and one that boasted distinguished alumni such as Nicolaus Copernicus and humanist and architect Leon Battista Alberti. Hoping to secure Clavius's recommendation, Galileo left with him a few of his original works on finding the center of gravity of various solids—a popular topic among Jesuit mathematicians at the time.

At about the same time, Galileo also proved an interesting

theorem that generated some buzz. He showed that if you take a
series of weights of, say, 1 libra (an ancient unit of weight equal to
approximately 11.5 ounces), 2 libras, 3 libras, 4 libras, and 5 libras,
and hang them at equal spaces along a balance arm, then the center
of gravity (the point around which the arm is in equilibrium)
divides the length of the balance arm precisely in a two-to-one
ratio. While this little theorem gained Galileo some recognition in
places ranging from Padua and Rome to universities in Belgium,
the chair at Bologna was still given to Giovanni Antonio Magini,
an established astronomer, cartographer (mapmaker), and mathematician from Padua.

This failure must have been a stinging blow to the young and
ambitious Galileo, but its impact was soon softened by a remarkable honor conferred upon him. In 1588 the consul of the Florentine
Academy, Baccio Valori, invited Galileo to deliver two lectures to the
academy on the geography and architecture of Dante's *Inferno* (hell)
in his masterpiece *The Divine Comedy*.

In this monumental poetic work (running more than fourteen
thousand lines), Dante tells the story of a poet's imaginative journey
through the afterlife, drawing inspiration from a wide range of philosophies. After an epic tour through the Inferno and Purgatory to
Paradise, the poet finally reaches that "love that moves the Sun and
other stars."

The invitation to present the lectures demonstrated the Academy's respect not only for Galileo's mathematical skills but also for
his literary scholarship. Galileo was undoubtedly delighted to receive
this request for two main reasons. First, mapping Dante's disorienting
description of hell in *The Divine Comedy* gave Galileo his first opportunity to attempt to build a bridge between a literary magnum opus
and scientific reasoning. In later years, an important part of what was
to become Galileo's continuous philosophy and ultimate legacy was
the demonstration that science is an integral part of culture, and that
it can enhance, rather than diminish, even the poetic experience. As
a means to this goal, he went against the long-standing tradition of

writing science in Latin and wrote instead in Italian. Working in the other direction, in his extensive scientific writing, Galileo drew from his literary resources to convey ideas and associations in a colorful, stimulating fashion.

Second, Galileo shrewdly recognized the importance of these lectures for his personal career. He was fundamentally asked to act as an arbiter between two contradictory commentaries and views on the location, structure, and dimensions of the *Inferno*, offered by two interpreters of Dante's work. One was the beloved Florentine architect and mathematician Antonio Manetti, biographer of the famous architect Filippo Brunelleschi. The other was the intellectual Alessandro Vellutello of Lucca. Vellutello argued that Manetti's amphitheater-like edifice could not be stable, and he offered an alternative model in which hell occupied a much smaller volume around the Earth's center. Much more than a purely highbrow dispute was at stake. Florence had suffered a humiliating military disaster at Lucca in 1430. After an unsuccessful besieging of that city, Brunelleschi, acting that time as an army engineer, came up with the idea of diverting the river Serchio, so as to surround Lucca with a lake and force it to surrender. The plan backfired cataclysmically when a dike failed, and the river flooded the camp of the Florentine army instead. This painful historical memory was surely on the minds of the members of the Florentine Academy when they asked Galileo to demonstrate that Manetti "had been slandered by Vellutello." Moreover, Vellutello's commentary represented a disowning of Manetti's authority—and, by association, the Florentine Academy's—in the interpretation of Dante. In other words, Galileo was entrusted with saving the academy's prestige, and he realized that by handing Manetti a victory over Vellutello, he could be regarded as a champion of Florentine pride.

Galileo started his first lecture with a direct reference to astronomical observations (probably having in mind the fact that most of the positions he was seeking at the time were in mathematics and astronomy) but emphasizing that deciphering the architecture of hell would require theoretical considerations. He then swiftly moved on to describing Manetti's interpretation, using the same analytical skills that would become his trademark in all of his scientific investigations. The dark scenery of Dante's hell occupied a cone-shaped portion of the Earth, with Jerusalem at the center of the cone's dome-shaped base and the cone's vertex being fixed at the Earth's center (Figure 2.1 shows Botticelli's depiction). Contrary to Vellutello's claim that Manetti's structure occupied a full one-sixth of the Earth's volume, Galileo used the geometry of solids he had learned from reading the works of Archimedes to demonstrate that, in fact, it filled less than seven-hundredths of the bulk—in his words: "less than one of the 14 parts of the whole aggregate." He then methodically proceeded to tear apart Vellutello's model by showing that not only would parts of his proposed architecture have collapsed under their own weight, but also that the design did not even agree with Dante's chilling description of the descent to hell. In contrast, Galileo argued that in Manetti's construction, "its thickness is sufficient . . . to sustain it." Galileo

Figure 2.1. Sandro Botticelli's *Chart of Hell*, based on Dante's *Inferno*.

finished his *Inferno* lectures by thanking the academy, to which he felt "most obligated," wisely adding that he thought he had demonstrated "how much subtler is the invention of Manetti."

Unfortunately, in perhaps wishing too much to please his audience, Galileo fell into his own trap. He didn't realize that Manetti's architectural structure was also prone to catastrophic collapse (not that any of his listeners noticed). Galileo may have discovered his blunder shortly after delivering the *Inferno* lectures, since he stopped referring to them for many years, and his biographer, Viviani, never even mentioned the lectures, despite having lived in Galileo's house during the master's last years.

Only in his last book, on the *Two New Sciences*, did Galileo return to the interesting problem of the strength and stability of constructions when they are scaled up in size. The key insight he had gained by then was that whereas the volume (and therefore the weight) increases a thousandfold when the size is blown up to be ten times larger, the resistance to cracking (which happens along two-dimensional surfaces) increases only a hundredfold, and therefore falls behind the increase in the weight. Galileo wrote in *Two New Sciences*: "The larger machine, made of the same material and the same proportions as the smaller one, in all other conditions will react with the right symmetry to the smaller one, except for its strength and its resistance against violent invasions; the bigger the ship, the weaker it will be." Then, most likely alluding to his *Inferno* mishap, he noted that "some time ago" he also had made a mistake when estimating the strength of scaled-up objects. Perhaps the most remarkable point about Galileo's flawed *Inferno* incident was the fact that even many years after having delivered a scientific talk about a poetic work, Galileo felt compelled to revisit his conclusions, revise his old ideas based on newly acquired acuity, and publish the new, correct results in an entirely different context for the problem.

Galileo was indeed a Renaissance man, but one may wonder whether in our age of narrowly focused specialization and career-driven attitudes such people still exist, and whether individuals who are curious about a wide range of topics or polymaths with broad interests are even needed. Drawing on about a hundred interviews with extraordinarily creative men and women across many disciplines, University of Chicago psychologist Mihaly Csikszentmihalyi suggested that the answer to both questions is in the affirmative. His conclusion: "If being a prodigy is not a requirement for later creativity, a more than usually keen curiosity about one's surroundings appears to be. Practically every individual who has made a novel contribution to a domain remembers feeling awe about the mysteries of life and has rich anecdotes to tell about efforts to solve them." Indeed, creativity often means the ability to borrow ideas from one field and transpose them into another. Charles Darwin, for example, took one of the pillars of his theory of evolution—gradualism—from his geologist friends. This was the notion that just as the surface of the Earth is shaped very slowly by the actions of water, the Sun, the wind, and geological activity, so too do evolutionary changes occur over hundreds of thousands of generations.

Recognizing that a shout-out for "Renaissance persons" can inspire creativity in the modern world does not mean giving up on specialization. With sources of information literally at our fingertips, even those ten thousand hours or so (that are supposedly required to become an expert in a given topic according to author Malcolm Gladwell, though this has been disputed by authors of the original study) can be shortened through more efficient learning practices and techniques. This time saving, combined with the fact that humans are living longer than ever before, means that there is nothing today (in principle, at least) to prevent people from being *both* experts and Renaissance persons.

Returning to Galileo's life, the reputation he had gained through his *Inferno* lectures and a strong recommendation he had eventually received from Clavius proved to be extremely fruitful. In the summer of 1589, Filippo Fantoni left the chair of mathematics at the University of Pisa, and Galileo, the former dropout from that university, was appointed to that post.

CHAPTER 3

A Leaning Tower
and Inclined Planes

Galileo's first appointment as a professor and chair of mathematics at Pisa lasted only from 1589 till 1592, yet one particular story associated with that period has generated an iconic image of Galileo. It is a picture of him dressed in his imposing academic gown, dropping balls of different weights from the top of the leaning tower of Pisa.

The original tale comes from Viviani, who in 1657 put together what he described as his recollections from a conversation he had with Galileo in the latter's final years:

A great many conclusions of Aristotle himself on the subject of motion were shown by him [Galileo] to be false which up to that time had been held as most clear and indubitable, as (among others) that speeds of unequal weights of the same material moving through the same medium did not at all preserve the ratio of their heaviness assigned to them by Aristotle, but rather, these all moved with equal speeds, he [Galileo] showing this by repeated experiments made from the height of the leaning tower of Pisa in the presence of other professors and all the students.

In other words, contrary to the view held by all Aristotelians that the heavier the ball, the faster it would fall, Viviani claimed that by dropping balls from the Leaning Tower (sometime between 1589 and 1592), Galileo had demonstrated that two balls of the same material but of different weights hit the ground simultaneously.

As if this story wasn't dramatic enough, later biographers and historians just kept adding more details that weren't included in Viviani's original account or in any other contemporary sources. For example, British astronomer and popularizer of science Richard Arman Gregory wrote in 1917 that members of the University of Pisa assembled at the foot of the Leaning Tower "one morning in the year 1591," even though Viviani never mentioned the precise year or the time of day. Gregory also added that one ball was "weighing a hundred times more than the other"—again a detail not given by Viviani. Author Francis Jameson Rowbotham, who wrote about the lives of great scientists, great musicians, great authors, and great artists, added in his 1918 vivid description that Galileo "invited the whole University to witness the experiment."

Others were equally inventive. Physicist and historian of science William Cecil Dampier Whetham tells us in 1929 that Galileo dropped "a ten-pound weight and a one-pound weight together," repeating the same values of weights that had been mentioned in an earlier biography by Galileo scholar John Joseph Fahie. All of these science historians and others took the tower story to mark a turning point in the history of science: a change from reliance on authority to experimental physics. The event had become so famous that in a fresco painted in 1816 by Tuscan painter Luigi Catani, Galileo is shown performing the experiment even in the presence of the grand duke. But did this demonstration really take place?

Most present-day historians of science think that it probably did not. The skepticism stems partly from Viviani's known tendency for ahistorical embellishments, partly from his occasional errors in recording the chronology of events, and perhaps mostly from the fact that Galileo himself never mentions this very specific experiment in his extensive writings, nor does it appear in any other contemporary documents. In particular, philosopher Jacopo Mazzoni, who was a professor at Pisa and a friend of Galileo, published a book in 1597, in which, while he generally supported Galileo's ideas on motion, he never mentioned an experiment by Galileo at the tower of Pisa. Similarly, Giorgio Coresio, a lecturer at Pisa who described in 1612 experiments that involved dropping objects from the top of the Tower of Pisa didn't attribute any of those to Galileo. We should note that Coresio made the strange claim that the experiments "confirmed the statement of Aristotle . . . that the larger body of the same material moves more swiftly than the smaller, and in proportion as the weight increases so does the velocity." This statement becomes especially puzzling when we realize that already in 1544, historian Benedetto Varchi mentioned experiments that had shown Aristotle's prediction to be wrong.

Galileo was seventy-five when Viviani came to live in his house; Viviani was eighteen, so embellishments could have come from both sides. I would argue, however, that from the perspective of appreciating Galileo's science, it is really not very important whether he performed this particular demonstration or not. The fact remains that during his years at Pisa, Galileo embarked on serious experimentation with free-falling bodies. This stands up irrespective of whether he dropped balls from the Leaning Tower or not. In Pisa, he also started composing a treatise analyzing various aspects of motion. This monograph, *De Motu Antiquiora* (*The Older Writings on Motion*), was published only in 1687, after Galileo's death, but its contents traces the development of his early ideas, and definitely puts Galileo (already during his early Pisan years) at the forefront of both experimental and theoretical investigations of motion in general, and of free-falling bodies in particular. In *De Motu* (*On Motion*) Galileo stated that he had confirmed by repeated experiments (without mentioning the Leaning Tower) that when two objects are dropped from a high place, the lighter one moves faster at first, but then the heavier object overtakes it and reaches the bottom first. This peculiar result has been shown by later experiments to be probably due to a nonsimultaneous release of the two objects. Basically, experiments have demonstrated that when each ball is held in one hand, the hand holding the heavier object becomes more tired, and it has to clasp the object with more force, resulting in a delayed release. Incidentally, the Flemish physicist Simon Stevin of Bruges dropped two balls of lead, one ten times the weight of the other, "from a point about 30 feet high," some years before Galileo's supposed Leaning Tower demonstration, and he published his results ("they landed so evenly that there seemed to be only one thump") in 1586.

De Motu marked the beginning of Galileo's serious criticism of Aristotle, and it formed the basis for his subsequent experiments with balls rolling down inclined planes. It also demonstrated that science sometimes progresses incrementally rather than as a result of revolutions. While Galileo's ideas about free-falling bodies departed significantly from those of earlier natural philosophers, at their initial stages they still did not quite square with the results of his experiments. The concepts inherited from Aristotle suggested that bodies fall at a constant speed, which is determined by the weight of the body and the resistance of the medium. To many, the fact that Aristotle had said so was sufficient to accept this as the truth. In *De Motu*, Galileo held that falling bodies accelerate (speed up), but only initially, and then they settle down to a constant proper speed, which is determined by the relative densities of the body and the medium. That is, he suggested that a ball made of lead moves faster [in Galileo's words, "far out in front"] than one made of wood, but that two balls of lead fall with the same speed, no matter how much they weigh. This was a step in the right direction, but not quite correct. For instance, Galileo realized that this description did not agree with the fact that free fall appeared to be continuously accelerating, but he thought that the speeding up itself might be gradually decreasing, eventually approaching a constant speed.

Only in his later book *Discourses and Mathematical Demonstrations Concerning Two New Sciences* (*Discorsi*) published in 1638, did Galileo arrive at a correct theory of free fall, according to which, in a vacuum, all bodies, irrespective of their weights or densities, *uniformly accelerate in precisely the same way.* Galileo put this explanation in the mouth of Salviati, Galileo's alter ego in the fictional dialogue in *Discorsi*: "Aristotle says, 'A hundred-pound iron ball falling from the height of a hundred braccia hits the ground before one of just one pound has descended a single braccio.' I say that they arrive at the same time." This crucial realization of Galileo's—the result of a vigorous experimental effort—was an essential prerequisite to Newton's theory of gravitation.

In modern times, in 1971 Apollo 15 astronaut David Scott dropped from the same height a hammer weighing 2.91 pounds and a feather weighing 0.066 pounds on the Moon (where there is virtually no air resistance), and the two objects struck the lunar surface simultaneously, just as Galileo had concluded centuries earlier.

Another problem with *De Motu* was that Galileo's early measurements, particularly of time, were still not sufficiently precise to allow for any definitive conclusions. He nevertheless had the foresight to make the following remark:

> When a person has discovered the truth about something and has established it with great effort, then, on viewing his discoveries more carefully, he often realizes that what he has taken such pains to find might have been perceived with the greatest ease. For truth has the property that it is not so deeply concealed as many have thought; indeed, its traces shine brightly in various places, and there are many paths by which it is approached.

In later years, questions such as "What is truth?" and "How is truth shown?" (especially in scientific theories) were to become essential in Galileo's life. These same questions have become perhaps even more critical today, when even indisputable facts are sometimes labeled "fake news." It is certainly true that, at their inception, the sciences were not immune to false beliefs, since they were sometimes connected to fictitious fields such as alchemy and astrology. This was partly the reason why Galileo decided later to rely on mathematics, which appeared to provide a more secure foundation. With the development of practices that allowed experiments to be reproduced (Galileo being one of the pioneers), scientific assertions have become increasingly more reliable. Basically, for a scientific theory to become accepted, even tentatively, it needs not only to agree with all the known experimental and observational facts, but also the theory must be able to make predictions, which can then be verified by subsequent observations or experiments. Not to accept the conclusions of studies that have passed all of these rigorous tests, with the associated uncertainties being clearly stated (as in climate change models, for instance), is tantamount to playing with fire—as is literally demonstrated by the weather extremes currently occurring around the globe and causing massive fires.

One of the impressions one might get from the immense efforts invested in his investigations of motion and the writing of *De Motu* is that Galileo had neglected his polymath roots and started to devote all his time to purely mathematical or experimental matters. This was definitely not the case. Even though Galileo spent much of his time in Pisa on empirical studies, his interest in philosophy and his love for poetry remained intact. In his writings, Galileo reveals an extraordinary grasp of Aristotle's teachings, notwithstanding the fact that sometimes he uses that very expertise to attack Aristotle's conclusions, as when he says: "How ridiculous is this opinion [Aristotle's] is clearer than daylight . . . if from a high tower two stones, one stone twice the size of the other, were flung simultaneously that when the smaller was halfway down the tower, the larger would have already reached the ground?" It is clear that Galileo did not absorb the knowledge and deep understanding of Aristotle just from drinking the Pisan water—he had to work hard to acquire it. In fact, as late as sixteen months before his death, Galileo still stated that he had consistently continued to follow Aristotle's logical methodology. In his own philosophy, however, Galileo stressed repeatedly the central role of mathematics. To him, true philosophy had to be a judicious mixture of observation, reasoning, and mathematics, with all three ingredients being absolutely necessary.

There was another amusing incident that happened at Pisa, in which Galileo demonstrated on one hand his admiration for the witty sixteenth-century poet Ludovico Ariosto (as well as for the parodic spirit of poet Francesco Berni), and on the other, his deep-seated aversion to authority and pompous formalism. It all started with a decree by the rector of the university requiring all professors to wear their academic gowns whenever they appeared in public. On top of the inconvenience imposed by this ridiculous order, Galileo was apparently further annoyed by the fact that he had been stiffly fined several times for breaking this rule. To express his disdain, he composed a 301-line satirical poem entitled "Capitolo Contro Il Portar la Toga" ("Against the Donning of the Gown"). In this rather risqué poem, Galileo reveals for the first time his contemptuous and provocative side and his inventive verbal humor—qualities he would make repeated use of in his later writings. In a few of the verses, he even advocates for people to walk naked because that would allow them to better appreciate one another's virtues. It is very likely that Galileo did not object just to the gown itself. Rather, he probably used the rule about the gown as a symbol for the dogmatic acceptance of Aristotle's authority by many of the scientists of his time. Alas, Galileo's mocking attitude most likely did not endear him to his Pisan colleagues. Here are a few lines from the controversial poem:

> *I shan't waste words but move out of my tower:*
> *Such fashion I will follow as is now in town,*
> *But how it pains and takes all my willpower!*
> *And pray don't think I'll ever don a gown*
> *As if I were a Pharisaic professor:*
> *I couldn't be convinced, not for a golden crown.*

Overall, Galileo managed to survive economically at Pisa, although his salary was a meager 60 scudi per year. This reflected the rather unglamorous status of mathematics at the time. By comparison, philosopher Jacopo Mazzoni was making more than ten times that amount at the same university. The death of Galileo's father in 1591 put an enormous financial burden on him, since he was the eldest son. He therefore sought, and fortunately obtained, an appointment at the University of Padua in 1592, where his salary was tripled. That prestigious chair had been vacant since the death of renowned mathematician Giuseppe Moletti in 1588, and university officials were rather picky in choosing a successor. Galileo's winning the position was aided greatly by the strong support of Neapolitan humanist Giovanni Vincenzo Pinelli, whose library in Padua—at the time the largest in Italy—functioned as an intellectual center, and whose strong recommendation carried enormous weight. Pinelli opened his library for Galileo, and it was there that he gained access to unpublished manuscripts and lecture notes on optics, all of which were to become helpful in Galileo's later work with the telescope.

Galileo would later describe his years in Padua—the city about which Shakespeare wrote: "fair Padua, nursery of the arts"—as the best time of his life. This was no doubt due largely to the freedom of thought and lively exchanges of information enjoyed by all scholars in the Venetian Republic, of which Padua was a part. These were also the years in which Galileo "converted" to Copernicanism.

PADUAN MECHANICS

Every researcher today knows that one cannot expect experimental results to demonstrate *precisely* any quantitative prediction. Statistical and systematic uncertainties (a range of values likely to enclose the real value)—creep into every measurement, making it sometimes difficult to even discern existing patterns at first glance. This concept runs contrary to the ancient Greeks' emphasis on very precise pronouncements. Living in a period in which no accurate measurements of time were possible, Galileo found the study of motion quite challenging and frustrating in his early attempts. In addition, his research was often interrupted by the fact that starting at about 1603, Galileo began to suffer from serious arthritic and rheumatic pains, which sometimes became so bad that they confined him to bed. These debilitating medical problems persisted, according to Galileo's son, "from about the fortieth year of his life to its end."

Nevertheless, from 1603 to 1609, Galileo developed a number of his ingenious methods for investigating motion, and a few of his groundbreaking results in mechanics had their roots in those years, too. Much later, in his book *Discorsi*, Galileo described both the problems he had faced in probing and analyzing the free fall of bodies, and his brilliant solutions. In particular, he had to overcome the seemingly insurmountable experimental difficulty of having to determine whether the speeds of objects of different weights were really equal or different after having been in free fall for relatively short time intervals. Galileo wrote:

> In a small height [from which the different bodies are dropped], it may be doubtful whether there is really no difference [in the speeds of the bodies or the precise times they hit the ground], or whether there is a difference, but it is unobservable. So I fell to thinking how one might many times repeat descents from small heights and accumulate many of those minimal differences of time that might intervene between the arrival of the heavy body at the terminus and that of the light one, so that added together in this way they would make up a time not only observable but easily observable.

This was already a remarkable insight. In an age that preceded the formulation of statistical methods, Galileo understood that if the same experiment is performed multiple times, the results can tease out and make credible even small differences. But his genius idea for these experiments was still to come. Galileo was seeking a way to literally slow down free fall, or to "dilute" gravity, so that the times of fall would be longer and more easily measurable, thereby making differences believable. Then it hit him: "I also thought of making moveables [objects] descend along an inclined plane not much raised above the horizontal. On this, no less than in the vertical, one may observe what is done by bodies differing in weight." In other words, a free-falling ball could be regarded as an extreme case of a ball rolling down an inclined plane when the plane is vertical. As Galileo's calculations show, by letting bodies slide (or roll) down an inclined plane tilted only by 1.7 degrees, he was able to considerably slow down the motion, to the point where he could make more reliable measurements.

In terms of his method for acquiring new knowledge, there is one interesting point that we should realize about Galileo's experiments in mechanics: his explorations were largely driven by theory or reasoning, rather than the other way around. In his words, from *De Motu*, one must "employ reasoning at all times rather than examples (for what we seek are causes of effects, and these causes are not given to us by experience)." About 350 years later, the great theoretical astrophysicist Arthur Eddington would echo a similar viewpoint: "Clearly a statement cannot be tested by observation unless it is an assertion about the results of the observation. Every item of physical knowledge must therefore be an assertion of what has been or would be the result of carrying out a specified observational procedure."

In Galileo's astronomical discoveries, on the other hand, observations were leading the way. Science progresses sometimes by experimental results preceding theoretical explanations, and sometimes by theories making predictions that are later confirmed (or falsified) experimentally or observationally. For example, it was known since 1859 that the orbit of the planet Mercury around the Sun did not quite agree with predictions based on Newton's theory of gravity. Einstein's theory of general relativity, which was published in 1915, *explained* the anomaly. At the same time, however, general relativity *predicted* that the path of the light from distant stars would be bent or deflected around the Sun to a certain degree. This prediction was first confirmed by observations made during a total solar eclipse in 1919, and it has since been reconfirmed by many subsequent observations. Arthur Eddington, by the way, led one of the teams that performed the observations in 1919.

The research concerning climate change today is progressing along similar steps. First, there has been an *observed* century-scale rise in the average temperature in the Earth's climate system. This was followed by studies aimed at identifying the main causes for this change, resulting in detailed climate models that have by now made predictions regarding the anticipated effects in the twenty-first century.

In spite of the fact that Galileo was happy in Padua on a personal level, this period marked a time of dire financial straits. His two sisters Virginia and Livia married in 1591 and 1601, respectively, and the obligation to pay the exorbitant dowries fell on Galileo. Worse yet, Virginia's husband even threatened that he would have Galileo arrested if he didn't pay the agreed-upon sum. While Galileo's brother Michelangelo also signed the dowry contract, he failed to make the payments, even though at the time he did succeed in securing two reasonable employments in succession as a musician. One of those was in Poland for which Galileo covered the travel expenses, and the other in Bavaria. As if to add insult to injury, while in Bavaria, Michelangelo married Anna Chiara Bandinelli and spent all of his money on an extravagant wedding banquet. Consequently, despite the fact that Galileo's salary at Padua climbed from its initial 180 scudi per year to 1,000 scudi by 1609, he constantly had to rely on giving private tutoring, accommodating about a dozen students for rent at his home, and selling instruments he manufactured at his workshop, to avoid becoming mired in serious debt. Casting occasional horoscopes for students and various socialites provided another source of much-needed income.

The fact that Galileo engaged in astrology is hardly surprising. One of the traditional functions of mathematicians at the time was to draw astrological charts. In addition, they were supposed to teach medical students how to use horoscopes to indicate the appropriate treatment. More than two dozen astrological charts drawn by Galileo have survived. Those include two charts for his own birth and charts for his daughters Virginia and Livia. However, we do know from a letter written by Ascanio Piccolomini, in whose house Galileo spent the first six months of his house arrest in 1633, that by that time the scientist derided astrology entirely and made fun of it "as a profession founded on the most uncertain, if not false, foundations."

The proximity of Padua to Venice allowed Galileo to forge new friendships and alliances with intellectuals and other influential figures there. One in particular, Gianfrancesco Sagredo, who owned a palace on Venice's Grand Canal, was to become almost like a brother to Galileo, and he was later immortalized in Galileo's *Dialogo*, playing the role of an intelligent and curious layperson. That depiction was apparently accurate, since in one of his letters, Sagredo gave the following self-evaluation of his own characteristics: "If I sometimes speculate about science, I do not presume to compete with the professors let alone criticize them, but only to refresh my mind by searching freely, without any obligation or attachment, the truth of any proposition that appeals to me." Another friend and close advisor, Fra Paolo Sarpi, in addition to being a prelate, a historian, and a theologian, was also a scientist and a superb mathematician with great interest in topics ranging from astronomy to anatomy. Galileo would later say in admiration: "No one in Europe comes before him [Sarpi] in knowledge of the [mathematical] sciences."

In 1608 Sarpi, who himself had an excellent command of optics and of the processes involved in vision, provided Galileo with the first reliable information about the invention of the telescope, after rumors of a Dutch device had spread throughout Europe. Even the polymath and playwright Giambattista della Porta confessed that he "never knew any man more learned" than Sarpi. This was the kind of praise reserved previously only for people such as Leonardo da Vinci, about whom King François I of France said that "he did not believe that a man had ever been born who knew as much as Leonardo."

Venice offered another important attraction to Galileo. Its celebrated arsenal—a complex of armories and shipyards—overflowed with instrumentation in which he showed great interest. It was said that at its peak, the thousands of men working at the arsenal could build a ship in one day. We shouldn't be surprised, therefore, that Galileo opened his book on the *Two New Sciences* with: "It seems to me that frequent visits to your famous Venetian arsenal open a large field of philosophizing on the part of speculative minds, especially in regard to the field in which mechanics is required. For every sort of instrument and machine is continually in use there by a large number of artisans." The fact that the space of the arsenal is used today for the Venice art Biennale acts as a symbolic reminder of the connection between art and science in Renaissance Italy.

All of these frenzied scientific and engineering activities in Venice's arsenal inspired Galileo to establish his own workshop, where he permanently employed an instrumentalist named Marcantonio Mazzoleni, who lived with his family in Galileo's house. The workshop (in some sense, a seventeenth-century equivalent of a start-up company today) served Galileo both in his own experimental investigations and in producing an income through the manufacturing of various measuring, surveying, and mathematical devices, some for military purposes. In particular, one such instrument, the geometric and military compass, was a type of calculator that aided in rapid computations of useful battlefield quantities such as the distance and height of a target. Galileo even published a small book in Italian (with only sixty copies distributed, to limit unauthorized access) demonstrating and explaining the operation of this calculator. Another scientist, Baldessar Capra, later published a book about the same apparatus, but in Latin, claiming falsely to have invented it—when in reality he had received lessons in its use from Galileo! Galileo's reaction was swift and aggressive. He collected affidavits from several people to whom he had demonstrated the instrument a few years earlier and accused Capra of plagiarism. After winning the case in front of university authorities, he followed up with a vicious article against Capra entitled "Defense Against the Calumnies and Impostures of Baldessar Capra."

Why did Galileo react so vehemently? There is little doubt that due to his economic struggles, he felt compelled to vigorously defend himself against any attack that could in any way tarnish his reputation and thereby diminish his chances of attaining a higher income or better employment opportunities. There was, however, probably an additional, personal honor element to Galileo's somewhat disproportionate reaction toward Capra. When a new star appeared in the sky in October 1604, Capra gloated in public that he had seen it five days before Galileo. This must have struck a nerve.

Galileo found more than mere intellectual and artistic stimulation in Venice. Through his friend Sagredo, he was introduced to the temptations that the Venetian night life had to offer—mainly fine wine and women. He formed a romantic association with Marina di Andrea Gamba, who consequently moved to Padua. The couple never married, but they stayed together for more than a decade and had two daughters, Virginia (later Sister Maria Celeste) and Livia (later Sister Arcangela), and a son, Vincenzo. One could speculate that Galileo's reluctance to enter into a formal marriage arrangement was influenced by the fact that the marriages in his immediate family were far from encouraging, but it is also possible that he gave up on a conventional relationship to be able to financially accommodate his sisters. At least, that's what his brother Michelangelo thought.

With regard to his scientific work, the most impressive results produced during his eighteen years in Padua came out of Galileo's experiments with inclined planes. Even though these results were not published until the 1630s, most of the experimental work was carried out in the period from 1602 to 1609. On October 16, 1604, Galileo wrote a letter to his friend Fra Paolo Sarpi in which he announced the discovery of the first mathematical law of motion—the law of free fall:

> Reconsidering the phenomena of motion . . . I can demon-
> strate . . . that the spaces passed over in natural motion [free fall]
> are *in proportion to the squares of the times* [emphasis added], and
> consequently the spaces passed over in equal times are as the odd
> numbers beginning from one. . . . Now the principle is this: that
> the body in natural motion increases its speed in the same pro-
> portion as its departure from the origin of its motion.

The first part of this statement is Galileo's discovered law: the distance that a free-falling body travels is proportional to the square of
the time of travel. That is, a body that free-falls for two seconds (from
rest) travels a distance that is four times (2 squared) longer than one
that free-falls for one second. In three seconds, a free-falling body
covers a distance that is nine times (3 squared) that of a body that falls
for one second, and so on. The second statement in Galileo's letter is,
in fact, a direct consequence of the first. Imagine that we'll call the
distance traveled during the first second of fall "1 Galileo"; then the
distance covered during the following one second will be the difference between 4 Galileos (the distance covered in two seconds) and 1
Galileo (the distance covered in the first second), which is 3 Galileos.
Similarly, the distance through which the body will fall in the third
second will be 9 Galileos minus 4 Galileos, which leaves 5 Galileos.
Consequently, the distances passed during successive periods of one
second will form the odd number sequence of 1, 3, 5, 7 . . . Galileos.

The last statement in the quote from Galileo's letter to Sarpi was
actually incorrect. In 1604 Galileo still thought that the speed of a
free-falling body increases in proportion to the *distance* from the
point from which the free fall had started. Only much later did Galileo realize that in free fall, the speed increases in direct proportion to
the *time* of fall and not to the distance. That is, the speed of an object
that has been free-falling for five seconds is five times the speed of
one that has been falling for only one second. In his later *Two New
Sciences*, he therefore correctly asserted: "Uniformly accelerated mo-

tion I call that to which, commencing from rest, equal velocities are added in equal times."

The importance of these discoveries for the history of science cannot be overemphasized. Whereas in the physics of Aristotle there were elements (such as earth and water) whose "natural motions" were supposed to be downward, Aristotle's theory contained also elements (such as fire) whose "natural motions" were upward, and air, whose natural motion depended on its location or surroundings. To Galileo, the only natural motion on Earth was downward (that is, toward the Earth's center), and it applied to all bodies. Entities that had been observed to float upward (such as air bubbles in water) did so only because of the lift force exerted on them by a medium of a higher density, as explained by the laws of hydrostatics originally formulated by Archimedes. We can recognize in these ideas some of the ingredients of Newton's theory of gravitation. What Galileo did not have an answer to—nor did he even try to answer—was *why* bodies fall at all. This was left for Newton. Galileo concentrated instead on discovering the "law," or what he regarded as the essence of free fall, rather than on causal explanations for free fall.

There was another aspect in which Galileo's ideas differed fundamentally from those of Aristotle. The Greek philosopher's theory of motion had never been put to any serious experimental tests, partly because of his (and Plato's) conviction that the correct way to discover truths about nature was to think about them rather than to perform experiments. To Aristotle, the only possible way to understand phenomena was to know their purpose. Galileo, on the other hand, employed a clever combination of experimentation and reasoning. He realized relatively early that progress is often achieved through correct decisions as to which questions should be asked, and also via studying artificial circumstances (as in the case of balls rolling down inclined planes) instead of examining only natural motions. This truly marked the birth of modern experimental physics.

Two elements in particular stand out as revolutionary in Galileo's new theory of motion: First, the universality of the law, which applies to all bodies in accelerated motion. Second, the extension of the formulation of mathematical laws from ones describing only static configurations that don't involve motion, as in Archimedes's law of the lever, to motion and dynamical situations.

A CONVERSION

Another facet of Galileo's years in Padua was the most significant for his future. Whereas many of his fruitful investigations were indeed in mechanics, the most important revision in his outlook on science was in astronomy. As noted earlier, in a publication entitled *Treatise on the Sphere, or Cosmography* (probably written in the late 1580s), Galileo still described and appeared to follow in detail the Ptolemaic geocentric system, without even mentioning the Copernican heliocentric model. This book probably reflected the requirements imposed by the university's curriculum, and it was used primarily for tutoring students. Two letters written in 1597, however, in which Galileo expresses for the first time his increasing conviction in Copernicanism, provide evidence for a radical change in his views.

The first letter, dated May 30, 1597, was addressed to Jacopo Mazzoni, a philosopher and former colleague of Galileo's at Pisa. Mazzoni had just published a book entitled *On Comparing Aristotle and Plato*, in which he argued that he had found proof that the Earth did not revolve around the Sun, thus invalidating the Copernican scenario. The argument was based on Aristotle's assertion that the top of Mount Caucasus, at the intersection of Europe and Asia, was illuminated by the Sun for a full one-third of the night. From this assumption, Mazzoni concluded incorrectly that since in the Copernican model an observer at the mountain's top (when the mountain was on the side of the Earth not facing the Sun) would be farther from the center of the world (the Sun), than in the Ptolemaic model (where the center of the world was assumed to be the Earth's center), the horizon of a Copernican observer should span much more than 180 degrees, contrary to experience. In his letter to Mazzoni, Galileo used precise trigonometry to show that the motion of the Earth around the Sun would not result in any detectable change in the visible portion of the celestial vault. Then, after refuting this presumed challenge to Copernicus, Galileo added a critical statement, saying that he "held [the Copernican model] to be much more probable than the opinion of Aristotle and Ptolemy."

Galileo's second letter was even clearer in expressing his views on Copernicanism. It came on the heels of a publication by Johannes Kepler. The great German astronomer is best remembered today for three laws of planetary motion that bear his name, which served as an impetus for Newton's theory of universal gravitation. Kepler was an accomplished mathematician, a speculative metaphysicist, and a prolific author. As a child, he was inspired by the spectacle offered by the comet of 1577. After studying mathematics and theology at the University of Tübingen, he was introduced to the theory of Copernicus by mathematician Michael Mästlin. Kepler seems to have been immediately convinced by the Copernican system, partly perhaps because the idea of a central Sun surrounded by the fixed stars with a space between the Sun and the stars appealed to his profound religious beliefs. He thought that the universe reflects its creator, with the unity of the Sun, the stars, and the intervening space symbolizing the Holy Trinity.

In 1596 Kepler published a book known as *Mysterium Cosmographicum* (*Cosmic Mystery*), in which he proposed that the structure of the solar system is based on the five regular solids known as the Platonic solids—the tetrahedron, the cube, the octahedron, the dodecahedron, and the icosahedron—being embedded one within the other. Since the five solids and the sphere of the fixed stars created precisely six spaces, Kepler thought that his model explained why there were six planets. (Only six were known at that time.) While the model itself was quite crazy, Kepler did assert in his book the Copernican view that all the planets orbited the Sun. His mistake was not in the details of the model but, rather, in his assumption that the number of planets and their orbits were some fundamental quantities that had to be explained from first principles. Today we know that the orbits of planets are just accidental results of the conditions that happened to prevail in the protosolar nebula.

The two copies of Kepler's book intended for astronomers in Italy somehow landed on Galileo's desk. On August 4, 1597, after having read only the preface, Galileo sent Kepler a letter in which he stated that he believed the Copernican model to be correct. He went even further, saying that he had been a Copernican "for several years," and adding that he found in the Copernican model a way to explain a number of natural phenomena that couldn't be explained by the geocentric scenario. But, he admitted, he had "dared not publish" any of those theories, as he had been deterred by the fact that Copernicus "appeared to be ridiculed and hissed off the stage."

In Kepler's reply of October 13, 1597, he urged Galileo to hasten and publish those explanations supporting the Copernican model, if not in Italy, then in Germany. Such a publication, however, didn't materialize. Since Galileo was never particularly shy or hesitant when it came to publishing what he regarded as solid results, this lack of action on his part suggests that at the time, before he made any observations with the telescope, Galileo might have had nothing more than hunches stimulated perhaps by his discoveries in mechanics. He was probably also thinking already about an explanation for sea tides, which he was later to develop into one of his main arguments for the motion of the Earth. As with the case of Mazzoni, those clues could have included Galileo's intuitive feeling that a few of the objections raised with regard to the motion of the Earth could be refuted. It is also possible, however, that Galileo's passiveness was political—a consequence of the fact that at that stage of his career, with Europe immersed in the Counter-Reformation era, he was somewhat reluctant to appear in Catholic Italy as an ally of Kepler, a known Lutheran.

An occurrence in the autumn of 1604 gave Galileo an opportunity to present publicly, if not quite a Copernican view, then at least a clear anti-Aristotelian position. On October 9 astronomers in a few Italian cities were startled to discover a new star—a nova—that rapidly became brighter than all the stars in the sky. Meteorologist Jan Brunowski observed it on October 10 and informed Kepler, who embarked on continuous, fruitful observations that lasted for almost a year. (That is why the object is known today as Kepler's supernova.) Baldessar Capra, who a few years later would have the dispute with Galileo over the compass/calculator, noticed the new star together with his tutor Simon Mayr and a friend Camillo Sasso, on October 10. The Italian friar and astronomer Ilario Altobelli informed Galileo, who first observed it in late October and then gave three lectures about the nova to huge audiences sometime between November and January. Galileo's main point was simple: since no displacement or shift had been observed in the position of the new star against the background of the distant stars—a phenomenon known as parallax—the star had to be farther than the Moon. However, that region, according to Aristotle, was supposed to be inviolable and immune to change. Therefore, the new star (which, by the way, we know today represented the explosive death of an old star, a phenomenon known as a supernova) contributed to shattering Aristotle's conception of an immutable stellar sphere.

That imaginary sphere had started cracking already in 1572, when the Danish astronomer Tycho Brahe discovered another "new" star—also an exploding, dying star known now as Tycho's Supernova. Somewhat unfortunately, perhaps, Galileo added another element to his "explanation" of the nova, which was completely wrong. He suggested that the new star represented a reflection of sunlight by "a large amount of vapor" ejected from Earth and projected past the Moon's orbit. If true, this would have dealt an even more fatal blow to the Aristotelian distinction between the degradable terrestrial matter and the eternally incorruptible celestial stuff, but in reality, this fanciful, supplemental idea was totally unnecessary, and Galileo himself had doubts about it.

Not everybody agreed that the new star all but destroyed the Aristotelian cosmos. It often takes more than one or two observations, regardless of how convincing they might be, to persuade people to abandon beliefs cherished for centuries. A few didn't even believe the nova to be located in Aristotle's presumed pristine celestial quintessence, mistrusting the parallax measurements. Others, such as the authoritative Jesuit mathematician and astronomer Christopher Clavius, confirmed the null parallax determination—that is, no shift having been observed—but refused to accept its implications as compelling. Still others, such as the Florentine philosopher Lodovico delle Colombe, with whom Galileo was to have grave disputes in later years, came up with alternative explanations for the nova's appearance. Wanting to preserve the incorruptibility of the heavens, delle Colombe suggested that the nova was not really a new star or an intrinsic change in the brightness of a star, but, rather, only a newly *observable* star. That is, a star becoming visible because of a swelling in the heavenly material acting like a lens. Galileo bothered to answer only a few of the critics, deeming the others unworthy of a response. In one case, his answer was cast in sarcastic dialogue form, which he composed with friends and published under a pseudonym.

Overall, the superb results in mechanics, the contemplation of new theoretical vistas in astronomy, and the artistic and free-spirited allure of Venice made life in Padua very enjoyable for Galileo. However, his financial troubles, which forced him to undertake a burdensome load of teaching, apparently weighed heavily on his mind. The difficulties and stress eventually caused him to start seeking better-paying opportunities with individual patrons as opposed to universities. Later, he candidly explained his motivation for moving from Padua in two letters written in 1609 and 1610:

> Greater leisure than I have here, I do not believe I could have elsewhere so long as I am forced to derive the sustenance of my household from public and private teachings. . . . To obtain any salary from a Republic, however splendid and generous, without rendering public service, is not possible, since to draw benefits from the public it is necessary to satisfy the public. In a word, I cannot hope for such benefits from anybody but an absolute ruler. . . . Hence . . .

I desire that the primary intention of His Highness shall be to give
me ease and leisure to bring my works to a conclusion without my
being occupied in teaching.

Galileo indeed moved to Florence in September 1610 at the invitation
of the Grand Duke Cosimo II de' Medici of Tuscany, but not before
he manufactured the instrument with which he was about to produce
his most breathtaking discoveries. His confidants in Venice regarded
the trading of intellectual freedom (which Galileo had in abundance
in Padua) for financial stability and release from teaching weariness
as a grave mistake. History has shown that even the long hand of the
Inquisition had rarely reached the Republic of Venice in any consequential way, while a move to Florence made Galileo vulnerable to
control by the Church. Knowing what we know today about Galileo's
fate, we therefore have to conclude that Galileo's Venetian friends
were absolutely right. Intellectual freedom is indeed invaluable. This
is especially important today, when truth and facts appear to be under
siege.

CHAPTER 4

A Copernican

If until 1609 Galileo's experiments concentrated on objects falling downward toward the Earth's center, starting that year he turned his attention upward to the heavens. Here is how that celestial adventure unfolded. In late 1608 Galileo's Venetian friend Paolo Sarpi heard a rumor about a spyglass—an optical gadget invented in the Netherlands—that could make distant objects appear closer and larger. Realizing that such an instrument could have interesting applications, Sarpi alerted Galileo in 1609. Around the same time, he also wrote to a friend in Paris to inquire whether the rumor was true.

In his publication *The Sidereal Messenger*, Galileo described the circumstances:

> About 10 months ago, a report reached my ears that a certain Fleming had constructed a spyglass by means of which visible objects, though very distant from the eye of the observer, were distinctly seen as if nearby. Of this truly remarkable effect, several experiences were related, to which some persons gave credence while others denied them. A few days later, the report was confirmed to me in a letter from a noble Frenchman at Paris, Jacques Badovere, which caused me to apply myself wholeheartedly to

investigate means by which I might arrive at the invention of a similar instrument. This I did soon afterwards, my basis being the doctrine of refraction.

The last sentence in this description could be a bit misleading, since it gives the impression that Galileo was guided by the theoretical principles of optics, a topic in which his knowledge was, in fact, rather scanty. In reality, Galileo's approach was much more experimental. He discovered through trial and error that by placing spectacle lenses in a tube, a plano-concave one at one end and a plano-convex one at the other, he could easily achieve a magnification of about three or four. Since Venice was an aspiring maritime power, Galileo immediately realized the bargaining potential that such a device (in his words: "of inestimable value in all business and every undertaking at sea or on land") could give him in salary negotiations with Venetian senators. He therefore swiftly embarked on learning how to polish higher quality lenses, and experimented with lenses of different sizes and distances apart. Amazingly, within less than three weeks, he was in Venice, equipped with an eight-powered telescope and, through Sarpi's connections, about to demonstrate this "perspicillum," as he called it, to Venetian decision-makers.

The ability to spot distant ships long before they could be seen with the naked eye sufficiently impressed the senators, who initially agreed to increase Galileo's salary from 520 to 1,000 scudi per year. To his disappointment, however, once the senate realized that the telescope was not an exclusive Galileo invention (even though he never claimed it to be), but, rather, a device already known elsewhere on the Continent, the salary increase was limited to one year, after which it was to be frozen. Furious about this turn of events, as well as about the fact that the senators did not seem to appreciate that his telescope was far superior to those that were circulating in Europe at the time, Galileo sent a telescope to the Grand Duke of Tuscany, Cosimo II de' Medici, in the hope of securing a court appointment in Florence. This might seem like a long shot on his part, but Galileo did have reasons for optimism. He had been Cosimo's tutor in mathematics during some summers between 1605 and 1608, and it was Cosimo's father, Ferdinando I de' Medici, who appointed Galileo to the professorship of mathematics at the University of Pisa in 1589.

Things started to progress in leaps and bounds at the end of 1609. In December of that year and January 1610 alone, Galileo probably made more earthshaking discoveries than any other person in the history of science. In addition, he managed to improve his telescope to fifteen-power by November 1609 and to twenty-power or more by March 1610. Turning this improved device to the night skies, he started by observing the surface of the Moon, moved on to resolving stars in the Milky Way, and then to making the revolutionary discovery of the satellites of Jupiter. Armed with these truly surprising findings, he decided to promptly publish his results, fearing that another astronomer might scoop him out of what he correctly perceived as momentous revelations. Indeed, *The Sidereal Messenger* (Figure 4.1) appeared in Venice as early as March 13, 1610. Perhaps not unexpectedly, Galileo's burst of creativity followed—and was almost certainly helped by—the departure of his mother from Padua. Not only did Giulia Ammannati not support her son in his research, she did no less than attempt to convince Galileo's servant, Alessandro Piersanti, to spy on his master. Constantly suspicious that Galileo's lover, Marina Gamba, would somehow persuade Galileo to reduce the financial support he was providing his mother, or that she would steal her linen, Giulia recruited Piersanti to secretly report to her about the couple's private conversations. If that wasn't enough, she even asked the servant to steal some of Galileo's telescope lenses, which she intended to give to her son-in-law, the husband of Galileo's sister Virginia, in what she perceived to be an act of gratitude for the son-in-law's alleged generosity. Fortunately, Piersanti immediately turned over Giulia's conspiratorial letters to Galileo.

Politically savvy at that point in his life, Galileo dedicated *The Sidereal Messenger* to Cosimo II de' Medici, Fourth Grand Duke of Tuscany. He went even further and named the four satellites of

Figure 4.1. Title page of *The Sidereal Messenger*.

Jupiter the Medicean Stars, because, he said, "the Maker of the stars himself admonished me to call these new planets by the illustrious name of Your Highness." The effect of these "heavenly" gifts was expeditious and gratifying. By June 1610, Galileo had been appointed philosopher and mathematician to the grand duke, and chief mathematician, free of teaching obligations, at the University of Pisa. When applying for the position, Galileo insisted upon having the title of "philosopher" added to his position of a court mathematician.

One reason for this request was simple: philosophers enjoyed a higher status than that of mathematicians. This was not, however, merely a status-affirming desire; Galileo confessed at the time that he had "studied more years in philosophy than months in mathematics."

Two of the distinguishing characteristics of many of those who truly made a difference in the history of science were: first, their ability to immediately recognize which discoveries could be genuinely impactful; and second, their effectiveness in disseminating their findings and in making them intelligible to others. Galileo was masterful at both. In 1610 he became unstoppable: in the course of just about one year, he discovered the phases of Venus, the fact that Saturn appeared to have a bizarre shape, and that variable spots move across the surface of the Sun. Over the succeeding couple of years, he also published two more books, *Discourse on Bodies in Water* in 1612, and *History and Demonstrations Concerning Sunspots* the following year.

The Sidereal Messenger became an instant best seller—its initial printing of 550 copies sold out in no time. Consequently, by 1611, Galileo became the most famous natural scientist in Europe. Even the Jesuit scientists in Rome had to take notice, and they rolled out the red carpet for him when he arrived for a visit on March 29. While the distinguished astronomer Christopher Clavius had some reservations about the interpretation of a few of the results, generally the mathematicians of the Collegio Romano expressed their trust in the accuracy of the observations themselves and certified the phenomena revealed by the telescope as real. As a result, Galileo was received in audience by Pope Paul V, and by Cardinal Maffeo Barberini, who years later (as Pope Urban VIII) played a crucial role in what has become known as the "Galileo affair." In addition, both Cardinal Roberto Bellarmino (often Anglicized as Robert Bellarmine), former rector of the Collegio, and Clavius himself, met with Galileo during his Roman visit, and Bellarmino even discussed some aspects of Copernican astronomy with him. The only sign of a potential cloud on the horizon came in the form of a somewhat ominous comment Bellarmino made to the Tuscan ambassador at the end of Galileo's stay in Rome: "If he [Galileo] had stayed here too much longer, they [church officials] could not have failed to come to some judgment upon his affairs."

Another honor conferred upon Galileo during the same trip was his election as a sixth member of Federico Cesi's Accademia dei Lincei (literally, the "Academy of the Lynx-Eyed"). This prestigious science academy had been founded in 1603 by Cesi, a Roman aristocrat (later Prince of Acquasparta), and three of his friends, and its idealistic goals had been stated as "not only to acquire knowledge of things and wisdom, and living together, justly and piously, but also peacefully to display them to men, orally and in writing, without any harm." It was named after both the sharp-eyed lynx and Lynceus, "the most keen-eyed of the Argonauts" in Greek mythology. The academy, its membership soon to grow even beyond the borders of Italy, published Galileo's book on sunspots in 1613, and later his book *The Assayer* in 1623. Galileo always felt very honored to be an academician, and he often signed his name as: "Galileo Galilei, Linceo." He and Cesi became bonded not only by their personal affinity, but also by their shared conviction that many beliefs held since antiquity about the natural world had to be abolished.

What precisely, then, were these wondrous observations of Galileo that for the first time showed humankind what the heavens were really like?

LIKE THE FACE OF THE EARTH ITSELF

In 1606 someone named Alimberto Mauri published a satirical book in which he speculated (based on reasoning inspired by naked-eye observations) that the features seen on the surface of the Moon indicated that the lunar surface was covered with mountains surrounded by flat plains. Many science historians suspect that Alimberto Mauri was really Galileo, writing under a pseudonym. Be that as it may, with the telescope in hand, Galileo finally had the opportunity to test this conjecture. Indeed, the Moon was the first celestial object to which he turned his spyglass. What he saw was a surface covered with blemishes and small circular areas that looked like craters. This was, however, where his artistic education came in handy. Observing in particular the terminator—the boundary separating the illuminated part from the dark one—and using his imaginative understanding of light and shadow and his grasp of perspective, Galileo was able to argue convincingly that the lunar terrain was very rugged. He described it as "uneven, rough, and crowded with depressions and bulges. And it is like the face of the Earth itself." Galileo's spectacular wash drawings and etchings (Figures 4.2 and 4.3) show points of light in the dark portion, which gradually increase in size toward the boundary.

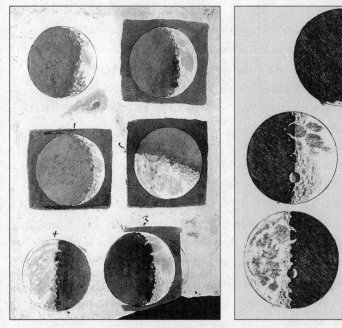 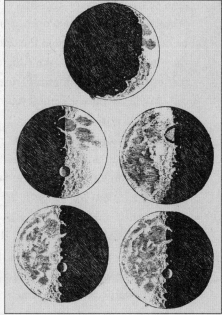

Figure 4.2. (*left*) Galileo's wash drawings of the Moon, as seen through his telescope.

Figure 4.3. (*right*) Galileo's etchings of the Moon.

This is precisely what one would expect when, at sunrise, only the tops of mountains are lit up, with the light later creeping down the mountains till it reaches the dark plains. Estimating the distance of one such point of light from the terminator to be about one-tenth of the Moon's radius, Galileo determined the height of that mountain to be more than four miles. The numerical value itself was later challenged in October 1610 by the German scientist Johann Georg Brengger, who suggested that ranges of mountains on the Moon probably overlap, or the Moon's rim would have appeared jagged rather than smooth. The precise height notwithstanding, Galileo demonstrated that he could not just see but also—in principle, at least—estimate fairly accurately the size of features in the lunar landscape. Today we know that the tallest mountain on the Moon is Mons Huygens, which is about 3.3 miles high. When we compare Galileo's drawings of the Moon's surface to images of the Moon taken with modern telescopes, it becomes immediately apparent that he deliberately exaggerated the dimensions of a few elements (such as the one known today as the Albategnius crater, shown in the bottom half of the lower etching in Figure 4.3), probably in order to didactically highlight the different levels of illumination and shadowing that he observed in the crater.

Galileo's drawings of the Moon provide us with yet another wonderful example of the overlap and interconnections between science and art in the late Renaissance. Somewhat surprisingly, in a famous painting entitled *The Flight to Egypt*, a German artist who worked in Rome at the time and died in December 1610, Adam Elsheimer, depicted the Moon in a fashion that is strikingly similar to Galileo's drawings. So much so, in fact, that a few art historians have even speculated that Elsheimer might have observed the Moon through one of the early telescopes, which could have been provided to him by his friend Federico Cesi.

An intriguing story related to the *Sidereus Nuncius* and art emerged in 2005, when an Italian art dealer named Marino Massimo De Caro offered to sell to the New York antiquarian Richard

Lan a remarkable copy of the *Sidereus Nuncius*. Instead of the usual etchings, this copy contained five wonderful watercolor drawings of the Moon, presumed to have been painted by Galileo himself. A battery of experts in the United States and Berlin confirmed the authenticity of the copy, which Lan bought for a half a million dollars. One of those experts, Horst Bredekamp, was so fascinated by the beauty of this specimen that he wrote a book about the exciting find. Then things took an unexpected turn. While writing a review of the English version of Bredekamp's book in 2011, Nick Wilding, a Renaissance historian at Georgia State University, started to suspect that something was not quite right with the new copy of the *Sidereus Nuncius*. To make a long story short, further examination and inquiry revealed that the copy was indeed a masterful forgery by the Italian seller De Caro.

Galileo used his lunar observations to discuss another puzzling topic that had generated many false interpretations over the years: the Moon's secondary light. Observers had been baffled by the fact that even the portions of the Moon that are dark when the Moon is at its crescent phase, are not pitch black—they appear to be dimly illuminated. In Galileo's words: "If we examine the matter more closely, we will see not only the extreme edge of the dark part shining with a faint brightness, but the entire surface . . . made white by some not inconsiderable light."

Previous explanations for this phenomenon ranged from the inconceivable suggestion that the Moon is partly transparent to sunlight, to the almost equally dubious proposition that the Moon doesn't just reflect sunlight but also shines with its own intrinsic light. Galileo readily dismissed all of these theories, calling some of them "so childish as to be unworthy of an answer." Then, even though he made it clear that "we will treat this matter at greater length in a book on the *System of the World*," he offered a brief explanation that was remarkable in its simplicity: just as the Moon provides some light to Earth at night, he argued, the Earth brightens the lunar night. This phenomenon is known today as earthshine. Probably sensing that this proposal might raise some objections among the Aristotelian faithful, Galileo quickly added a clarifying touch:

> What is so surprising about that? In an equal and grateful ex-change, the Earth pays back the Moon with light equal to that which she received from the Moon almost all the time in the deepest darkness of the night. . . . In this sequence [of lunar phases], then, in alternate succession, the lunar light bestows upon us her monthly illuminations, now brighter, now weaker. But the favor is repaid by the Earth in like manner.

A beautiful photograph of the lit Earth rising above the lunar horizon was taken from lunar orbit by Apollo 8 astronaut Bill Anders on December 24, 1968 (Figure 6 in the color insert). Because of the Moon's synchronous rotation with its orbital motion (the same side of the Moon always faces Earth), such an Earthrise can be seen only by an observer in motion relative to the lunar surface.

Galileo finished the discussion of his sweeping discoveries about the Moon with a powerful proclamation:

> We will say more in our *System of the World*, where with very many arguments and experiments a very strong reflection of solar light from the Earth is demonstrated to those who claim that the Earth is to be excluded from the dance of the stars [planets], especially because she is devoid of motion and light. For we shall demonstrate that *she is movable and surpasses the Moon in brightness* [emphasis added], and that she is not the dump heap of the filth and dregs of the universe.

Even though Galileo did not analyze the full implications of his lunar findings in *The Sidereal Messenger*—this was left for his *Dia-logo*—what could be inferred from them was fairly transparent. First, according to the Aristotelian cosmology (which over the centuries had become intertwined intimately with Christian orthodoxy), there was a clear distinction between things terrestrial and things celestial. Whereas everything on Earth was corruptible, mutable, could be eroded, decay, or even die, the heavens were supposed to be perfect, pure, enduring, and immutable. Unlike the four classical elements that were supposed to be the constituents of everything earthly— earth, water, air, and fire—heavenly bodies were believed to be made of a fifth, different, immaculate substance dubbed "quintessence," or literally the fifth essence. Yet Galileo's observations showed that there were mountains and craters on the Moon and that by reflecting the Sun's light, the Earth behaved very much like any other planetary object. No proof was given at this stage for the suggestion that the Earth was really moving, but Galileo's declaration that "she is movable" spoke volumes toward Copernicanism. If the Moon was, in fact, solid and very much like Earth, and it moved in an orbit around the Earth, why couldn't the Earth, which was Moonlike, move around the Sun?

Understandably, this new picture of the lunar surface and the Moon's place in the world provoked vehement objections. After all, it stood in stark contrast to the surreal description in the book of Revelation: "A great and wondrous sign appeared in the heaven: a woman clothed with the sun, with the moon under her feet and a crown of twelve stars on her head." Traditionally, in artistic portrayals of this biblical description, the Moon had been represented by a perfectly smooth, blemish-free, translucent object, symbolizing the Virgin's perfection and purity and continuing the Greek and Roman mythology of the personification of the Moon as a goddess. But Galileo's lunar deviation from prevalent convictions was only the beginning. His other discoveries with the telescope were about to deliver the coup de grace to the old cosmology.

STARRY, STARRY NIGHT

After the Moon, Galileo directed his telescope to those other points of light that shine brightly in the night's sky—the stars—and there, too, a few surprises awaited. First, unlike the Moon (and later the planets), the stars did not appear any larger through the telescope than they did to the naked eye, although they seemed brighter. From this fact alone, Galileo concluded correctly that the apparent sizes of the stars when observed with the unaided eye were not real but, rather, just artifacts. He did not know, however, that the apparent sizes were actually caused by the stellar light being scattered and refracted in the Earth's atmosphere rather than by anything related to the stars themselves. Consequently, he thought that the telescope removed the stars' misleading "adventitious irradiation." Nonetheless, since he couldn't make out the images of stars with the telescope, Galileo deduced that the stars were much farther from us than the planets.

Second, Galileo discovered scores of faint stars that couldn't be seen at all without the telescope. For example, in close proximity to the Orion constellation, he counted no fewer than five hundred stars, and he found tens more close to the six most brilliant Pleiades stars. Even more consequential for the future of astrophysics was Galileo's discovery that stars varied enormously in brightness, with some being a few hundred times brighter than others. About three centuries later, astronomers created diagrams in which the stellar luminosity was displayed against the stellar color, and the patterns observed in those diagrams have led to the realization that the stars themselves evolve. They are born out of clouds of gas and dust, they spend their lives generating power through nuclear reactions, and they die, sometimes explosively, after running out of energy sources. In some sense, this could be regarded as the final nail in the coffin of the Aristotelian notion of unchanging heavens. Still, the most surprising result concerning stars came when Galileo pointed the telescope to the Milky Way. That apparently smooth, luminous, and mysterious band across the sky broke up into countless faint stars packed closely in clusters.

These findings had significant implications for the Copernican-Ptolemaic debate. Some years earlier, the famous Danish astronomer Tycho Brahe pointed out what he perceived as a serious difficulty for the heliocentric theory. If the Earth was really orbiting around the Sun, he contended, then in observations taken six months apart (when the Earth is at two diametrically opposed points along its orbit), the stars should have shown a detectable displacement in position—a parallax—against the background, in the same way that trees observed through the window of a moving train appear to move with respect to the horizon. For such a shift not to be detected, Brahe argued, necessitated that the stars should be at very great distances. However, one could then estimate the size that the stars had to be in order for them to be seen with their apparent dimensions observed with the naked eye. Those expanses turned out to be even larger than the diameter of the entire solar system, which seemed highly implausible. Consequently, Brahe concluded that the Earth could not be orbiting the Sun. Instead, he proposed a revised, hybrid geocentric-heliocentric system in which all the other planets orbited the Sun, but the Sun itself orbited the Earth.

Galileo had a much simpler explanation for the absence of parallaxes. As we saw earlier, he concluded that the apparent dimensions of stars as seen with the naked eye did not represent real physical sizes—they were just artifacts. Stars were indeed at such large distances, he claimed, that shifts in their positions could not be detected, even with the then-available telescopes. Galileo was right: the detection of parallaxes had to await the development of higher-resolution telescopes. A stellar parallax was first observed only in 1806 by the Italian astronomer Giuseppe Calandrelli. The first successful measurements of a parallax were made by the German astronomer Friedrich Wilhelm Bessel in 1838. As of 2019, the European Space Agency's Gaia space observatory, launched in 2013, has determined the parallaxes to more than a billion stars in the Milky Way and nearby galaxies.

Galileo also rejected Brahe's "compromise" solar system model for two main reasons: First, the model appeared to him to be extremely contrived—in today's language, it had "too many moving parts." Second, in later years, Galileo relied on a moving Earth to explain the phenomenon of sea tides (discussed in chapter 7). He therefore could no longer accept a scenario in which the Earth had to be at rest. Galileo's intuition on rejecting a hybrid model proved correct.

The picture that emerged from Galileo's observations of the stars was very different from Aristotle's ancient concepts. Rather than being studded on a solid celestial sphere positioned just beyond the orbit of Saturn, stars were now much smaller in apparent size than previously thought, numerous beyond counting, and spanning huge ranges in both brightness and distance. In fact, this star system was beginning to look dangerously similar to the speculative cosmos depicted by mathematician and philosopher Giordano Bruno, in which multiple worlds existed in an infinite universe. Being well aware of Bruno's tragic end—he was burned alive on February 17, 1600—Galileo was very careful in describing and interpreting his observations of the stars, even in his later book the *Dialogo*. Careful or not, however, Galileo's observations of distant stars in the Milky Way galaxy can definitely be regarded as humanity's first peek at the vast universe that exists beyond the solar system.

Today we know that the Milky Way contains between 100 billion and 400 billion stars. Based primarily on data from the Kepler and Gaia Space Telescopes, recent estimates of the number of roughly Earth-sized planets in the Milky Way, which are orbiting Sun-like stars in that just right, not-too-hot and not-too-cold "Goldilocks" zone (known as the habitable zone) that allows for liquid water to exist on the planetary surface, put it in the billions.

A COURT HAS BEEN FOUND FOR JUPITER

On the evening of January 7, 1610, Galileo observed the planet Jupiter through his twenty-power telescope and noticed that, in his words, "three little stars were positioned near him—small but yet very bright." Galileo added that these stars intrigued him "because they appeared to be arranged exactly along a straight line and parallel to the ecliptic." Two of the stars were on the east of Jupiter and one on the west. The following night, he again saw the three stars, but this time they were all west of the planet and equally spaced, which made him think that perhaps Jupiter was moving to the east, contrary to expectations based on the astronomical tables that existed at the time.

Clouds prevented Galileo from observing on the ninth, but on the tenth he saw only two stars to the east. Guessing that the third star was hidden behind Jupiter, he was beginning to suspect that Jupiter was not moving much after all; rather, it was those stars that were moving. This celestial dance continued, with only two stars appearing on the east on January 11, and the third star reappearing on the west (and two on the east) on the twelfth. On the thirteenth, a fourth star appeared (three on the west and one on the east), and on January 15 all four were on the west. (It was cloudy again on the fourteenth.)

The earliest surviving record of Galileo's observations of Jupiter and what turned out to be its satellites is on the bottom half of a draft letter to the Doge of Venice (Figure 7 in the color insert). That page is now in the Special Collections of the University of Michigan—Ann Arbor. Fascinatingly, Galileo's drawings in the document reveal that at least until January 12, it had not occurred to him that the satellites could be orbiting Jupiter. Rather, he assumed that the three objects were moving along a straight line, in a very non-Copernican fashion. When the fourth satellite appeared on the thirteenth, however, Galileo realized that his assumption could not be correct, since it required one satellite to literally pass through another. Only after the fifteenth did the correct explanation dawn upon him. The conclusion from these meticulous observations now seemed inescapable:

> Since they sometimes follow and at other times precede Jupiter by similar intervals, and are removed from him toward the east as well as the west by only very narrow limits, and accompanying him equally in retrograde and direct motion, no one can doubt that they complete their revolutions about him while, in the meantime, all together they complete a 12-year period about the center of the world [referring to Jupiter's orbit around the Sun].

In plain English, Galileo discovered that Jupiter had four satellites, or moons, orbiting it, and, like our Moon, they were revolving approximately in the same plane as other planetary orbits. Jupiter was exhibiting a miniature Copernican system. On January 30 he informed the Tuscan state secretary, Belisario Vinta, that the four satellites move around a larger "star" (planet) "like Venus and Mercury, and perhaps other known planets, do around the Sun." He confirmed this fact through diligent, methodical observations of the satellites on every clear night until March 2. During this period, he also determined the distances of the moons from Jupiter and from one another and measured their brightness. To convince everybody else of what he saw, he presented no fewer than sixty-five diagrams showing the different configurations of satellites he had observed.

The discovery of Jupiter's four satellites was not only of historical significance—these were the first new bodies revealed in the solar system since antiquity—it also demolished one of the serious objections to the heliocentric model. Aristotelians maintained that the Earth could not keep possession of its Moon while orbiting the Sun. They also raised a legitimate question: If the Earth is a planet, why is it the only planet to have a Moon? Galileo decisively silenced both of these demurrals by showing that Jupiter—which was clearly moving, since it was orbiting either the Sun (in the Copernican view) or the Earth (in the Ptolemaic system)—was still able to retain not just one but four orbiting moons! He put it very clearly in *The Sidereal Messenger*:

> We have moreover an excellent and splendid argument for taking away the scruples of those who, while tolerating with equanimity the revolution of the planets around the Sun in the Copernican system, are so disturbed by the attendance of one Moon around the Earth while the two together complete the annual orb around the Sun that they conclude that this constitution of the universe must be overthrown as impossible. For here we have only one planet revolving around another while both run through a great

circle around the Sun: but our version offers us four stars wandering around Jupiter like the Moon around the Earth while all
together with Jupiter traverse a great circle around the Sun in the
space of 12 years.

After the publication of *The Sidereal Messenger*, Galileo continued to
observe Jupiter's satellites for almost three more years, until he was
satisfied that he had accurately determined the periods of their revolutions around Jupiter. He referred to this gigantic observational and
intellectual endeavor as an "Atlantic labor," alluding to Atlas, who was
ordered by the god Zeus to support the sky on his shoulders. Even
the great astronomer Johannes Kepler had believed it impossible to
determine the periods, since he saw no obvious way to unambiguously identify and distinguish among the three inner moons. Astonishingly, Galileo's results for the periods agree with modern values to
within less than a few minutes.

Today there are seventy-nine known moons of Jupiter (fifty-three
are named and the others awaiting official names). Eight of these are
believed to have formed in orbit about the planet, and the others were
probably captured. Out of the four Galilean satellites, as they are
now called, two, Europa and Ganymede (the latter of which is larger
than the planet Mercury), are each thought to contain a large ocean
underneath a thick crust of ice. Both moons are considered potential
candidates for harboring simple life forms under the ice, a fact that
no doubt would have delighted Galileo. The innermost of the four
Galilean satellites, Io, is the most geologically active body in the solar
system, with more than four hundred known active volcanoes. The
fourth Galilean moon, Callisto, is the second largest of the four.

What would have undoubtedly annoyed Galileo no end is the fact
that the Galilean satellites are known today by the names assigned
to them by the German astronomer Simon Mayr rather than as the
"Medici stars." Mayr may have independently detected the satellites
before Galileo, but he failed to understand that the moons were orbiting the planet. Galileo regarded Mayr as a "poisonous reptile" and

an "enemy not only of me, but of the entire human race," ever since he convinced himself that Mayr was the villain behind Baldessar Capra's plagiarizing the geometric and military compass. Galileo wrote about Mayr that while in Padua (where Galileo resided at the time), "he set forth in Latin the use of the said compass of mine, appropriating it to himself, had one of his pupils [Capra] print this under his name. Forthwith, perhaps to escape punishment, he departed immediately for his native land [Germany], leaving his pupil in the lurch, as the saying goes."

FOR THIS OLD MAN, TWO ATTENDANTS

The detection of Jupiter's four satellites was the last of Galileo's world-changing discoveries made in Padua. It only whetted his appetite for more breakthroughs. No wonder, then, that soon after moving to Florence, he aimed his telescope at the next giant planet in terms of its distance from the Sun: Saturn. The initial observations were disappointing, however, since they did not reveal any satellites. This situation changed when, on July 25, 1610, his inspection revealed something that looked like two motionless stars, one attached to Saturn on each side. To avoid being scooped, and still at a stage where he was making discoveries faster than he could publish them, Galileo sent a jumbled series of letters announcing his discovery to Kepler via the Tuscan ambassador to Prague. This was common practice at the time, to establish priority of discovery using a riddle, thereby not actually disclosing what had been found. Galileo's coded description was:

smaismrmilmepoetaleumibunenugttauiras.

Kepler had initially failed to make any sense of Galileo's message, and the English astronomer and Kepler's correspondent Thomas Harriot fared no better. But from the fact that the Earth had one Moon and Jupiter four moons, he concluded that Mars had to have two moons, so as to form the geometric progression 1, 2, 4, and so on. Guided by this mathematical belief and assuming that Galileo had discovered the satellites of Mars, Kepler eventually succeeded in creating a message out of Galileo's string of letters that differed by only one character from the original jumble: *Salve umbistineum geminatum Martia proles*, meaning roughly: "Be greeted, twin companionship, children of Mars."

Ingenious as Kepler's solution was, it had nothing to do with Galileo's intentions. The decoded message in the string of letters was supposed to read: *Altissimum planetam tergeminum observavi*, which translates into: "I have observed the highest of the planets [Saturn] three-formed."

On November 13, 1610, Galileo finally revealed precisely what he meant:

> I have observed that Saturn is not a single star but three together, which always touch each other, they do not move in the least among themselves and have the following shape oOo. . . . If we look at them with a telescope of weak magnification, the three stars do not appear very distinctly and Saturn seems elongated like an olive. . . . A court has been found for Jupiter, and now for this old man two attendants who help him walk and never leave his side.

To Galileo's amazement, those apparently reliable "attendants" completely disappeared by late 1612. He expressed his bewilderment in a letter to his correspondent, German humanist, historian, and publisher Markus Welser: "Were the two smaller stars consumed like spots on the Sun? Have they suddenly vanished and fled? Or has Saturn devoured his own children?" Despite his puzzlement, Galileo dared to predict that those "stars" would reappear in 1613, which they did, this time resembling ears or handles, one on each side of Saturn.

While Galileo was able to correctly predict again, in 1616, another vanishing of the "handles" ten years later, his prediction was apparently based on the assumption that these were similar to Jupiter's moons. A true explanation for the strange ears had to wait until the 1650s, when Dutch mathematician and astronomer Christiaan Huygens identified them as the now-famous Saturn's rings. Since the rings are flat and relatively thin, they were essentially undetectable when seen edge-on, and appeared like ears when their surface was inclined at a larger angle to the line of sight or when seen face-on.

It is interesting that we now know that the rings didn't always exist, nor will they last forever. The rings are estimated to be no older than about 100 million years, a short time compared with the approximately 4.6-billion-year-old solar system. More surprisingly, however, a study published in December 2018 found that due to "ring rain"—the draining away of the rings onto the planet in the form of a dusty rain of ice particles—the rings will disappear in about 300 million years. Galileo and we are therefore lucky to have lived in the relatively "short" period of time that allowed us to see this spectacular phenomenon.

Amusingly, Kepler's anticipation of Mars having two moons turned out to be correct, even though this had nothing to do with a geometric series. Furthermore, in his famous 1726 satire *Gulliver's Travels*, the English writer Jonathan Swift (who may have been inspired by Kepler) wrote about two Martian moons. In 1877 American astronomer Asaph Hall discovered the moons, now called Phobos and Deimos.

THE MOTHER OF LOVE

One of the major objections raised against the heliocentric model had to do with the appearance of the planet Venus. In the Ptolemaic geocentric model, Venus is always more or less between the Earth and the Sun, so it was expected always to appear as a crescent of varying dimensions (but never as much as half full, Figure 4.4a). In the Copernican model, on the other hand, since Venus was assumed to orbit the Sun, and it is closer to the Sun than the Earth, it was expected to exhibit the complete series of phases like the Moon, appearing fully illuminated as a small, bright disk, when farthest from the Earth and as a dark, large one, when closest (and as a large crescent just before that; Figure 4.4b). Through a series of painstaking observations between October and December 1610, Galileo definitively confirmed the predictions of the Copernican model. His decision to embark on these observations (and the interpretation of the results) may have been inspired and certainly encouraged by a detailed letter he had received from Benedetto Castelli, in which Castelli empha-

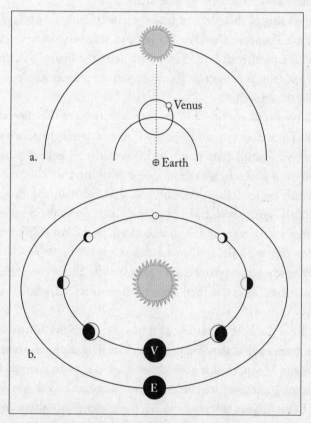

Figure 4.4. The expected appearance of Venus in the
Ptolemaic model (a) and in the Copernican model (b).

sized the significance of observing the phases of Venus. This was the first clear indication of the superiority of the Copernican view over the Ptolemaic one.

On December 11 Galileo hurried to send Kepler another mysterious anagram: *Haec immatura a me jam frustra leguntur oy*, meaning roughly "This was already tried by me in vain too early." (The use of the word *oy* has humorously been taken to hint at a potential Jewish ancestry for Galileo.) Frustrated by his inability to solve the puzzle, Kepler wrote to Galileo: "I adjure you not to leave us long in doubt of the meaning. For you see you are dealing with real Germans. Think in what distress you place me by your silence."

Responding to this plea, on January 1, 1611, Galileo sent Kepler the unscrambled version: *Cynthiae figuras aemulatur mater amorum*, which meant: "The mother of love [Venus] emulates the figures of Cynthia." The Greek female name *Cynthia* was used sometimes as a personification of the Moon.

Galileo was so confident in his interpretations of the observations that on December 30, 1610 he had sent a letter to Christopher Clavius—who until that point had objected to Copernicanism on both physical and religious grounds—explaining that all the planets shined only by reflecting sunlight, and that "without any doubt, the center of the great revolutions of all the planets" was the Sun.

Galileo's statements didn't go unnoticed. In the last edition of Clavius's comments on the influential astronomy textbook *The Sphere*, by the thirteenth-century astronomer Johannes de Sacrobosco, the Jesuit mathematician, who was then seventy-three years old, admitted:

Far from the least important of things seen with this instrument [the telescope] is that Venus receives its light from the Sun as does the Moon, so that sometimes it appears to be more like a crescent, sometimes less, according to its distance from the Sun. At Rome I have observed this, in the presence of others, more than once. Saturn has joined to it two small stars, one on the east, the other on the west. Finally, Jupiter has four roving stars, which vary their places in a remarkable way both among themselves and with respect to Jupiter—as Galileo Galilei carefully and accurately describes.

As he realized that these results all but shattered the Ptolemaic model, Clavius then added cautiously: "Since things are thus, astronomers ought to consider how the celestial orbs may be arranged in order to save these phenomena."

Galileo's observations of the phases of Venus dealt a fatal blow to the Ptolemaic geocentric systems, but they could not definitively dispose of Brahe's geocentric-heliocentric compromise, in which Venus and all the other planets revolved around the Sun, while the Sun itself orbited the Earth. This left a potential escape route for those Jesuit astronomers who were still determined to avoid Copernicanism.

I LIKEN THE SUNSPOTS TO CLOUDS OR SMOKES

Galileo did not neglect the object that, in the Copernican view, was central and most important in the solar system—the Sun itself—and his observations had led to the detection and first coherent explanation of the relatively dark areas of the Sun's surface known as sunspots. Galileo was definitely not the first person to discover these mysterious spots. Chinese and Korean astronomers may have seen them centuries earlier; for example, there are records from China that date back to the Han dynasty, 206 BCE to 220 CE. Sunspots had certainly been talked about at the time of Charlemagne, who ruled much of western Europe in the late-eighth to early-ninth centuries, and the Italian poet and painter Raffael Gualterotti described a sunspot he had seen on December 25, 1604, in a book published the following year. The English mathematician and astronomer Thomas Harriot observed sunspots with a telescope in December 1610 but didn't publish his results. His observations became known only in 1784 and were not published until 1833. Finally, the Frisian (northwest

Germany) astronomer Johannes Fabricius observed sunspots with a telescope on February 27, 1611, and described them in a pamphlet (of which Galileo was unaware) published the same year in the city of Wittenberg.

Observations of the Sun with the telescope were tricky, since without covering the lens with some protective material, such as soot, one could easily be blinded. Fortunately for Galileo, his talented former student Benedetto Castelli came up with the clever idea of simply projecting the image of the Sun formed by the telescope onto a screen or a sheet of paper. Originally, Galileo described his observations of sunspots in the preface to his book *Bodies in Water*. At that stage, he still entertained two possibilities for the nature of the spots. He thought that they could be either directly on the Sun's surface (in which case their motion indicated that the Sun rotated about its axis), or they could be planets revolving around the Sun very close to its surface. By the time of the book's second printing in the fall of 1612, however, Galileo was convinced that the spots had to be on the Sun's surface and "carried around by rotation of the Sun itself," and he added a paragraph to that effect.

The findings of another astronomer forced Galileo to turn his full attention to sunspots. In March 1612 he received from his German correspondent Markus Welser three letters describing sunspots, written under the pseudonym *"Apelles latens post tabulam"* ("Apelles hiding behind the painting"). The letters, later published as a pamphlet, had been written by the Jesuit priest and astronomer Christoph Scheiner, a professor at the University of Ingolstadt. Scheiner was forbidden from publishing under his real name for fear that if he happened to be wrong, the publication could discredit the Jesuits. Consequently, he used a pseudonym coined after the fourth-century Greek artist who used to hide behind his paintings to listen to viewers' criticism, implying that Scheiner was waiting for comments before revealing his identity. The astronomer maintained that the spots were projected shadows of many small planets orbiting the Sun in very tight orbits.

While there is no doubt that his ideas were inspired mainly by an attempt to rescue the Sun from imperfection, Scheiner based his model on three main arguments: First, the spots did not return to the same points, which to him implied that they were not contiguous to the surface of a rotating Sun. Second, Scheiner believed that the spots were darker than the unilluminated portions of the lunar surface, which he thought to be impossible if they were really on the Sun's surface. Third, the spots appeared thinner near the Sun's edge than when they were near the center of the solar disk, which he interpreted as an example of phases, like the ones seen in the case of Venus.

In addition to his comments about sunspots, Scheiner called attention to what he regarded as more convincing evidence than the phases, that Venus was really orbiting the Sun. Scheiner's proof relied on the fact that tables based on the Ptolemaic model, known as ephemerides, predicted that Venus would transit (that is, pass in front of, as seen from Earth) the Sun on December 11, 1611. Yet no such long-duration transit had been observed.

Welser sent the letters to Galileo to ask for the famous scientist's opinion on Scheiner's ideas, apparently assuming that Galileo would appreciate the scientific approach exhibited in the letters. However, the answer to "Apelles" that he received from Galileo was quite different from what he had expected. On one hand, Galileo's reply was witty, quite courteous, and certainly scientifically brilliant, but on the other, it also used highly critical and rather patronizing language. For example, referring to what he regarded as Scheiner's obstinate adherence to some Aristotelian concepts (such as the hardness and immutability of the Sun), Galileo wrote that Apelles "cannot yet totally free himself from those fancies previously impressed on him."

Galileo's response was delivered in several installments. First, he sent two letters written in Italian ("because I must have everyone able to read it") in May and October. Then, after Scheiner replied to the first letter with a letter of his own, and Welser published the entire series of Scheiner's letters under the title *A More Accurate Disquisition of Sunspots and the Stars Wandering Around Jupiter*, Galileo dispatched a third letter in December. These three letters were also published in Rome by the Lincean Academy in March 1613, entitled *History and Demonstrations Concerning Sunspots and Their Phenomena* (Figure 4.5).

Galileo, never known for taking criticism well, was particularly annoyed by Scheiner's claim that the failure to detect Venus's transit constituted superior evidence that Venus orbited the Sun. He pointed out that Scheiner erred in estimating the planet's size and, in addition, that it would have sufficed for Venus to possess just a tiny bit

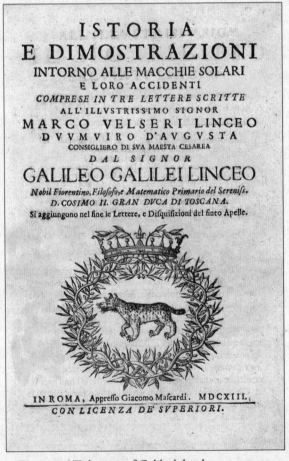

Figure 4.5, Title page of Galileo's book on sunspots.

of intrinsic brightness to render the absence of the transit useless in terms of proof for Venus's orbit.

Following this captiousness, Galileo moved on to dismantling Scheiner's explanation for sunspots. He made it clear that the spots were not really dark; they only appeared dark relative to the Sun's bright disk but were, in fact, brighter than the surface of a full Moon. He then argued correctly that the fact that the spots moved at varying speeds and changed positions with respect to one another showed unambiguously that they couldn't be satellites, since "anyone who wished to maintain that the spots were congeries of minute stars would have to introduce into the sky innumerable movements, tumultuous, uneven, and without any regularity." Instead, Galileo placed the spots squarely on the Sun's surface or no farther from the surface of the Sun than clouds would be (relatively) from Earth. Like clouds, he commented, the spots appeared suddenly, changed shape, and disappeared without warning. Using an intuition gained through his artistic education in drawing, Galileo also demonstrated that the spots' apparent narrowing as they approached the edge of the solar disk was due simply to the foreshortening that is observed when something moves on the surface of a sphere (Figure 4.6). Finally, and

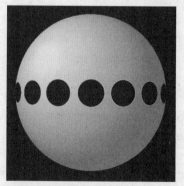

Figure 4.6. The visual phenomenon of foreshortening circles
drawn on the surface of a sphere.

perhaps most important, from the motion of the spots, Galileo estimated that the Sun takes approximately a month to rotate about its
axis. Indeed, we know now that the solar rotation period at the equa-
tor is 24.47 days.

Materially for Galileo's problems with the Catholic Church in
later years, he also asked Cardinal Carlo Conti for his opinion on
sunspots. The cardinal answered in July 1612 that there was nothing
in Scripture to support the Aristotelian notion of an incorruptible
Sun. Concerning Copernicanism in general, however, Conti advised
that this theory was inconsistent with Scripture, and that a different
interpretation of the biblical language "should not be admitted unless
it is really necessary."

Galileo's observations and interpretation of sunspots were of
critical significance for two main reasons. First, they demonstrated
that a celestial object could be spinning around its axis without either slowing down or leaving behind cloudlike features. This could
instantly remove two serious objections raised against the idea of a
spinning Earth in the Copernican model. The deniers asked: How
can the Earth keep spinning? And: Why aren't clouds (or birds, for
that matter) falling behind? Second, by publishing his results on the
rotating Sun in *Bodies in Water*—a book ostensibly about floating
bodies—Galileo signaled the first appearance of a *unified* theory of
the physics of the Earth and of the heavens. This type of unification
would later help produce Newton's universal theory of gravitation
(which brought together phenomena as diverse as apples falling
on Earth and planets orbiting the Sun), and would inspire all the
attempts today to formulate a "theory of everything"—a framework
merging all the fundamental interactions (electromagnetic, strong
and weak nuclear, and gravitational).

As he had often done, Galileo used the opportunity of the correspondence about sunspots to offer a glimpse into his philosophy
with respect to disseminating knowledge. In a letter to his friend
Paolo Gualdo, the archpriest at the Padua Cathedral, he made a few
remarkable comments about the fact that science should not be ex-

clusively the province of scientists. He explained that he hoped that from his letters to Welser, even those who "became convinced that in those 'big books there are great new things of logic and philosophy and still more that is over their heads'" would see that "just as nature has given them, as well as the philosophers, eyes with which to see her works, so she has also given them brains capable of penetrating and understanding them." Here Galileo establishes himself firmly as a member of what author John Brockman dubbed the "third culture": a direct conduit between the scientific world and laypeople. The key point that Galileo made was that scientific knowledge, when presented adequately, is not beyond the grasp of nonscientists, and since he regarded it as an essential part of human culture, literally everybody should strive to acquire it.

Fascinatingly, Galileo expressed here even less surprise at the human capacity to fathom the cosmos than Einstein did in 1936: "The eternal mystery of the world is its comprehensibility. . . . The fact that it is comprehensible is a miracle." Galileo's comments on the human ability to decipher nature's secrets were also echoed in his famous *Letter to Benedetto Castelli*, when he said that he did not believe "that the same God who has given us our senses, reason, and intelligence wished us to abandon their use."

Today we know that sunspots are indeed regions on the surface of the Sun that are somewhat cooler (temperature of about 4,000 Kelvin) than the surrounding area (about 6,000 Kelvin), and therefore appear darker. The lower temperature results from concentrations of magnetic field flux that suppress heat transport by convection (fluid motion). Sunspots typically last anywhere from a few days to a few months, and their sizes vary widely, from a few tens of miles across to a hundred thousand miles. Sunspot activity cycles last about eleven years; over the course of a cycle, the number of spots initially increases rapidly and then declines more slowly.

Galileo's *Letters on Sunspots* not only gave him a scientific victory over Christoph Scheiner at the time, but also brought Copernicanism to the attention of a larger readership. In 1615 Scheiner sent Galileo another work, entitled *Sol Ellipticus (The Elliptical Sun)*, and asked for Galileo's opinion on it, but he never received a response. Scheiner himself eventually published in 1630 an impressive and authoritative book on sunspots, which, in honor of his protector, Prince Paolo Orsini, he entitled *Orsini's Rose, or the Sun's Variations in Accordance with the Observed Appearance of Its Flares and Sunspots*. In this book, Scheiner conceded that the spots were on the solar surface, but he claimed that Galileo's conclusions on this topic had not been based on scientific reasons. Unfortunately, there is no doubt that Galileo's rather disparaging letters, his disregard for Scheiner's work in 1615, and some further comments he made later in his book *The Assayer*, which the Jesuit astronomer took to be directed at him personally, did turn Scheiner into an unappeasable enemy. This marked just the beginning of a conflict with the Jesuits, which would culminate in the punitive actions against Galileo in 1633.

CHAPTER 5

Every Action Has a Reaction

Given the magnitude of Galileo's celestial discoveries with the telescope and the fact that *Sidereus Nuncius* quickly turned him into an international celebrity, it was only to be expected that the reactions would be intense, passionate—and mixed. Indeed, controversy erupted almost before the ink dried on the book's pages. There were several reasons for the initial skepticism, and those could be traced to the long-lasting dominance of Aristotle's ideas and the almost religious acceptance of his general approach to science.

First, Galileo's methodology introduced a radically new element into what he asserted could be considered as constituting evidence. Fundamentally, Galileo claimed that his new device—the telescope—was revealing unimaginable truths that couldn't be perceived by the unaided senses. This flew in the face of the established Aristotelian tradition. How could anyone be sure that what Galileo was seeing was a genuine heavenly phenomenon and not a spurious artifact produced by the telescope itself? The telescope was, after all, the very first gadget presented as a means to boost and expand the power of sensory faculty.

A second problem that Galileo's discoveries in both mechanics and astronomy encountered had to do with his proclamation that the universe was "written in the language of mathematics." That is, he introduced the mathematization of the physical world. This notion ran completely contrary to Aristotelian reasoning, according to which mathematics had little if anything to do with reality or with the makeup of the cosmos. Until Galileo's time, astronomers were expected to use mathematics only to calculate planetary orbits and the apparent motion of the Sun, and thereby to create maps of the sky at particular times. These, in turn, were supposed to aid in estimating time, in establishing a calendar, in navigation, and in the production of astrological charts. Astronomers were not meant to construct physical models of the universe or of any phenomena within it. Here is how the Aristotelian Giorgio Coresio—the person who claimed that balls dropped from the Tower of Pisa confirmed Aristotle's assertions about free-falling bodies—had put it: "Let us conclude, therefore, that he who does not want to work in darkness must consult Aristotle, the excellent interpreter of nature." Compare this submissiveness to authority to Galileo's almost poetic later pronouncement in *The Assayer*: "It [the universe] is written in the language of mathematics, and the characters are triangles, circles, and other geometrical figures, without which it is humanly impossible to comprehend a single word of it, and without which one wanders in vain through a dark labyrinth."

Vincenzo di Grazia, a professor at Pisa, expressed his views on what he regarded as the contradistinction between mathematics and the natural sciences in even stronger terms:

Before we consider Galileo's demonstrations, it seems necessary to prove how far from the truth are those who wish to prove natural facts by means of mathematical reasoning, among whom, if I am not mistaken, is Galileo. All the sciences and all the arts have their own principles and their own causes by means of which they demonstrate the special properties of their own object. *It follows that we are not allowed to use the principles of one science to prove the properties of another* [emphasis added]. Therefore, anyone who thinks that he can prove natural properties with mathematical argument is simply demented, for the two sciences are very different.

Galileo could not have disagreed more with this attempt at hermetic compartmentalization of the different branches of science. "As if geometry in our day was an obstacle to the acquisition of true philosophy; as if it were impossible to be a geometer as well as a philosopher, so that we must infer as a necessary consequence that anyone who knows geometry cannot know physics, and cannot reason about and deal with physical matters physically! . . . as if knowledge of surgery was opposed to medicine and destroyed it," he mocked di Grazia. Einstein would fully agree with Galileo more than three centuries later, writing: "We may in fact regard [geometry] as the most ancient branch of physics. . . . Without it, I would have been unable to formulate the theory of relativity."

These two problems—the legitimacy of the telescope as an instrument enhancing the senses on one hand, and the role of mathematics in revealing truths about nature on the other—combined in the minds of the Aristotelians to form what they considered to be a powerful argument against Galileo's findings. Not only wasn't there a convincing theory of optics that could demonstrate that the telescope doesn't deceive, they contended, but also the validity of such a theory in itself, being based on mathematics, was questionable. On top of these philosophical matters, weighed, of course, the fact that all of Galileo's celestial discoveries defied Aristotelian ideas that the conservative establishment had revered for almost two millennia.

No wonder, then, that the immediate reaction in many circles was one of confusion. People from all ranks and spheres, ranging from state rulers and high church officials, to the lay public, turned to prominent scientists for opinion and advice. Even the German scholar Markus Welser, who later was instrumental in helping to spread Galileo's ideas, wrote to Christopher Clavius at the Collegio Romano, asking for his judgment:

With this occasion, I cannot neglect to tell you that it has been written to me from Padua as a certain and secure thing that with a new instrument called by many *visorio*, of which he makes himself the creator, Mr. Galileo Galilei of that university has discovered four planets, new to us, having never been seen, as far as we know, by a mortal, and also many fixed stars, not known or seen before, and marvelous things about the Milky Way. I know very well that "to believe slowly is the strength of wisdom," and I have not made up my mind about anything. I ask Your Reverence, however, candidly to tell me your opinion about this fact in confidence.

Another person who immediately understood the value of Clavius's support for the discoveries was Galileo's friend the painter Cigoli. Having had the impression that Clavius regarded the discovery of Jupiter's satellites as a hoax, he urged Galileo to visit Rome as soon as possible. This was sound advice, since Clavius was not the only person of authority in Rome who was skeptical. Christoph Grienberger, an Austrian Jesuit astronomer who eventually succeeded Clavius as professor of mathematics at the Collegio Romano, also suggested initially that Galileo's mountains on the Moon were nothing more than fanciful imaginations and that Jupiter's moons were merely optical illusions.

In the spring of 1610, many others were still equally incredulous. A fellow Florentine working in Venice at the time, Giovanni Bartoli, wrote on March 27: "They [the science professors] laugh at [these discoveries], and call them rash, while he [Galileo] tried to make them a great feat, and he has done so, and gained an increase in salary of 500 fiorentini." Bartoli added that many of those professors "think that he [Galileo] has made fun of them when he gave out as a secret the common spyglass which is on sale in the street for four or five lire, of the same quality, it is said, as his."

One more problem that Galileo faced was technical. Most of the telescopes circulating in Europe were either of very poor quality or difficult to use, and often both. This predicament was compounded by the fact that even with proper instruction, some people simply failed to see the phenomena that Galileo was claiming to have observed. As an example, when Galileo stopped at the University of Bologna on his way back to Padua from his meeting with the Grand Duke in Florence, he tried to demonstrate his findings to the chief astronomer there, Giovanni Antonio Magini—who, in 1588, had beaten Galileo to that post. Unfortunately, neither Magini nor anyone from his entourage succeeded in seeing Jupiter's satellites, even though in Galileo's own observation log, he recorded that on those two nights, April 25 and 26, he saw two and four moons, respectively.

Much worse yet, a Bohemian mathematician, Martin Horky, who at the time was working as an assistant to Magini and was even living in his home, wrote to Kepler a vicious letter describing Galileo's visit, in which he stated scornfully: "Galileo Galilei, the mathematician of Padua, came to us in Bologna, and he brought with him that spyglass through which he sees four fictitious planets." Horky added that he "tested that instrument of Galileo's in innumerable ways" and that while "on Earth it works miracles, in the heavens it deceives, for other fixed stars appear double." Horky went on to say that "most excellent men and most noble doctors" including philosopher Antonio Rofféni, "acknowledged that the instrument deceived." Superfluously and brutally, Horky even included a description of Galileo's physique, which, while probably partly accurate, given Galileo's continuous battle with poor health, was also vitriolic: "His hair hung down; his skin, in its tiniest folds, is covered with marks of the *mal français* [syphilis]; his skull is affected, delirium fills his mind; his optic nerves are destroyed because he has scrutinized minutes and seconds around Jupiter with too much curiosity and presumption. . . . [H]is heart palpitates because he has sold everyone a celestial fable."

Horky finished with a sentence in German, which, perhaps more than anything, revealed his treacherous character: "Unknown to anyone, I have made an impression of the spyglass in wax, and when God aids me in returning home, I want to make a much better spyglass than Galileo's."

As could perhaps be expected, Horky's ambitions never material-ized. His burning jealousy and fiery hatred toward Galileo, however, were not extinguished. In June he published in Modena, Italy, a tract entitled *A Very Brief Pilgrimage Against the Sidereal Messenger Recently Sent to All Philosophers and Mathematicians by Galileo Galilei*, which was really nothing more than a vicious diatribe against Galileo. Horky sought to deny the reality of Galileo's discoveries, but his ar-guments were laughable. He further stated maliciously that the sole purpose of the points of light Galileo claimed to have seen near Jupi-ter was to satisfy Galileo's lust for money.

While this particular incident backfired in a big way and ended well for Galileo—disgusted by Horky's actions, Magini evicted him from his home, and Kepler had nothing to do with him anymore—Horky's publication was symptomatic of the general reaction of adre-nalized Aristotelian devotees.

AN OUTREACH CAMPAIGN

Galileo knew right away that he had a difficult persuasion battle on his hands, but he never shied away from polemics, and he was prepared to fight for what he strongly believed to be true. First and foremost, he had to convince his former pupil and future employer, the Grand Duke Cosimo II de' Medici himself. To achieve this goal, he first dazzled the duke by showing him the spectacular views of the Moon through the telescope, probably as early as 1609. Later, he made sure that the duke would receive a high-quality telescope with detailed instructions on how to use it, as soon as *The Sidereal Mes-senger* hit the press in March 1610. Consequently, by the end of April, Galileo already knew that he could count on the duke's support. He then had to consider whom it was most advantageous to win over next. Realizing shrewdly that he who pays the piper calls the tune, he decided to reach out to the *patrons* of scientists rather than to the scientists themselves. Accordingly, Galileo sketched out an incredibly ambitious outreach plan to the Tuscan court:

In order to maintain and increase the renown of these discoveries, it appears to me necessary . . . to have the truth seen and
recognized . . . by as many people as possible. I have done and
am doing this in Venice and Padua. [Galileo indeed gave three
successful public lectures about his discoveries in Padua.] But
spyglasses that are most exquisite and capable of showing all the
observations are very rare, and among the sixty that I have made,
at great cost and effort, I have been able to find only a very small
number. [In fact, in the spring of 1610, he managed to have acceptable lenses for no more than about ten telescopes.] These
few, however, I have planned to send to great princes, and in
particular to the relatives of the Most Serene Grand Duke. And
already I have been asked by the Most Serene Duke of Bavaria
[Maximilian I, who employed Galileo's brother Michelangelo as a
lutenist] and the elector of Cologne [Ernest of Bavaria], and also
by the Most Illustrious and Reverend Cardinal Del Monte [an
important Venetian patron of Galileo], to whom I shall send [a
spyglass] as soon as possible, together with the treatise. My desire
would be to send them also to France, Spain, Poland, Austria,
Mantua, Modena, Urbino, and wherever else it would please His
Most Serene Highness.

A few others who, for obvious reasons, were on Galileo's list of early
recipients of the book and/or a telescope were various cardinals such
as Scipione Borghese, who in addition to being a great patron of the
arts was also the nephew of Pope Paul V, and Odoardo Farnese, another cardinal patron of the arts and the son of the Duke of Parma.
Oddly enough, but not inconsistent with his character, Galileo's focus
was primarily on the success of his outreach program; so much so
that he did not include his own brother Michelangelo among those
who were to receive a telescope.

Fortunately for Galileo, the grand duke supported unequivocally
the promotion efforts. Not only did the Tuscan court finance the
manufacturing of all the necessary spyglasses, but also Tuscan ambas-

sadors in the major European capitals were sent copies of *The Sidereal Messenger* and were charged with the task of helping to further Galileo's discoveries. Why did the Medicis provide such indefatigable assistance to Galileo? Not because of their interest in the Copernican model, but because they recognized Galileo's unusual ability and talent in presenting his discoveries as emblems of Medici power.

The endeavors started to bear fruit in April 1610. On April 19 Johannes Kepler, the most distinguished European astronomer at the time, delivered his ringing endorsement of Galileo's findings. Amazingly, while Kepler had already read Galileo's book, he offered his blessing and approval even before he had a chance to confirm the discoveries through his own observations: "I may perhaps seem rash in accepting your claims so readily with no support of my own experience," he wrote. "But why should I not believe a most learned mathematician, whose very style attests to the soundness of his judgment?" Then, in stark contrast to Horky's charges of deceit, Kepler continued: "He has no intention of practicing deception in a bid for vulgar publicity, nor does he pretend to have seen what he has not seen." Finally, describing the essential characteristics of a truly great scientist, Kepler pronounced: "Because he loves the truth, he does not hesitate to oppose even the most familiar opinion, and to bear the jeers of the crowd with equanimity."

With respect to the observations themselves, Kepler made various speculations about the findings, some of them quite farfetched. For instance, he suggested that there were living beings on the Moon who constructed some of the observed features. In addition, since Kepler shared the prevalent religious belief that all cosmic phenomena must have a purpose, he reached the following imaginative deduction: "The conclusion is quite clear. Our Moon exists for us on the Earth, not for the other globes. Those four little moons exist for Jupiter, not for us. Each planet in turn, together with its occupants, is served by its own satellites. From this line of reasoning, we deduce with the highest degree of probability that Jupiter is inhabited." Kepler did not know that Jupiter is a gas giant with no solid crust. The best chances for any form of life in the Jovian system are, in fact, on a couple of the moons.

Not all of Kepler's inferences were so fanciful. For example, when discussing the fact that the fixed stars and the planets appeared differently when observed with the telescope, he made the following, astonishingly prescient remark: "What other conclusion shall we draw from this difference, Galileo, than that the fixed stars generate their light from within, whereas the planets, being opaque, are illuminated from without; that is, to use [philosopher Giordano] Bruno's terms, the former are suns, the latter moons or earths?"

Today we indeed make the clear distinction between stars, which generate their own luminosity through internal nuclear reactions, and planets, which primarily reflect the light of their host stars.

In May 1610 Kepler published his letter under the title *Dissertatio cum Nuncio Sidereo (Conversation with the Sidereal Messenger)*. Since Galileo was clearly pleased with its content, the letter was reprinted in Florence later in the year. At that point, praise started pouring in from all directions. Galileo was hailed as a Columbus of the heavens. The Scottish librarian Thomas Segeth raved: "Columbus gave man lands to conquer by bloodshed, Galileo new worlds harmful to none. Which is better?" Sir Henry Wotton, an English diplomat in Venice who managed to lay his hands on one of the very first scarce copies of *The Sidereal Messenger*, sent it to King James I of England on March 13, 1610, accompanied by a letter that read, in part: "I send herewith unto his Majesty the strangest piece of news . . . that he hath ever yet received from my part of the world; which is the annexed book . . . of the Mathematical Professor of Padua, who by the help of an optical instrument . . . hath discovered four new planets rolling about the sphere of Jupiter, beside many other unknown fixed stars."

Another Englishman, astronomer Sir William Lower, who managed to hear about the discoveries in southwest Wales (testifying to the success of the outreach effort), sent on June 11, 1610, an even more enthusiastic letter to astronomer Thomas Harriot, saying: "Me thinkes my diligent Galileus hath done more in his threefold discoveries [referring to the mountains on the Moon, resolving the stars in the Milky Way, and Jupiter's satellites] than Magellane [Portuguese explorer Ferdinand Magellan] in opening the streights to the South Sea or the duchmen that weare eaten by beares in Nova Zembla." He was referring here to the Dutch navigator Willem Barentsz and his crew, who were stranded on the Arctic archipelago of Nova Zembla in 1596–97 while searching for a Northeast passage.

In France, at a commemoration held in the Loire Valley for the late King Henry IV on June 6, 1611, students recited a poem entitled "Sonnet on the Death of King Henry the Great, and on the Discovery of Some New Planets, or Stars Wandering Around Jupiter, Made This Year by Galileo Galilei, Famous Mathematician of the Grand Duke of Florence." The king, stabbed to death a year earlier by a religious fanatic, had indeed shown keen interest in Galileo's work, but never got to see the discoveries with his own eyes. His widow, Queen Marie de' Medici (by then regent for her son, King Louis XIII), sent word to Florence requesting one of "Galilei's large spyglasses." Unfortunately, the first instrument delivered to her was not of a particularly high quality, reflecting Galileo's difficulty in producing superior telescopes. Only in August 1611 was Galileo able to furnish the queen with an adequate spyglass, winning instant admiration. The grand ducal ambassador Matteo Botti wrote from France:

> Having presented to Her Majesty the Queen your instrument, I showed her that it is much better than another one sent earlier. . . . Her Majesty liked it very much, and she even kneeled on the ground, in my presence, to see the Moon better. She enjoyed it infinitely and was very pleased by the compliment I offered her in your name, which was accompanied by much further praise, not only on my part but also on Her Majesty's who demonstrates that she knows and admires you, as you deserve.

In Italy, the Medici commissioned poems about the discoveries from a number of Jesuit poets. Some of the excessively saccharine ones compared Galileo to Atlas, whose prowess forces even the heavens to turn on new stars. Venetian poet and glassmaker Girolamo Magagnati also wrote a few verses, in a pamphlet entitled *A Poetic Meditation upon the Medici's Planets*, conveying the glorious merits of Galileo's discoveries:

> *But you, O Galileo of the Ether, crossed*
> *Boundless inaccessible fields,*
> *And sank the curious plow*
> *Of errant spirit into the eternal sapphires,*
> *Turning over the Sky's golden clouds*
> *You discovered new Orbs and new Lights.*

Perhaps the most impressive tribute was provided by Galileo's friend the painter Cigoli, who was commissioned by Pope Paul V to create a fresco for the dome in the Pauline Chapel in the church of Santa Maria Maggiore. *Assumption of the Virgin*, executed between September 1610 and October 1612, depicted the Virgin standing on the Moon. The amazing element in this fresco was that Cigoli painted the Moon not as a smooth, spotless sphere, but rather precisely as it looked in Galileo's drawings of what he saw through the telescope (Figure 5 in the color insert).

BELIEVE IN SCIENCE

In his epic poem *The Aeneid*, Virgil wrote: "Believe one who has proved it. Believe an expert." Indeed, at some point, expert, professional confirmations of Galileo's observations and findings started coming in from other astronomers, and once that happened, at least the validity of what this new Columbus of the night's sky saw could no longer be questioned or disputed. During September 1610, both Kepler in Prague and Antonio Santini, a Venetian merchant and amateur astronomer, saw Jupiter's satellites. Kepler used the telescope that Galileo had sent to Ernest of Bavaria, the elector-archbishop of

Cologne, and Santini used a homemade telescope. Later in the fall, astronomer Thomas Harriot in England and astronomers Joseph Gaultier de La Valette and Nicolas-Claude Fabri de Peiresc in France also detected those four Medici moons. Astronomer Simon Mayr discovered them independently in Germany.

There remained the critically important opinion of the astronomers of the Collegio Romano, and of Clavius in particular. As late as October 1, 1610, Galileo's friend Cigoli reported: "Clavius said to one of my friends about the four stars [Jupiter's satellites] that he laughs at them, and that it will be necessary to make a spyglass which produces them and then show them, and that Galileo can keep his opinion, and he will keep his." However, as the story of the new discoveries gained momentum and became a hot topic of discussion all across Europe, church officials could not fail to take notice of the potential implications for the Church's orthodoxy. Consequently, the head of the Collegio Romano and chief theologian of the Holy Office (responsible for defending Catholic doctrine), Cardinal Roberto Bellarmino, asked the Jesuit mathematicians to specifically either confirm or refute five of Galileo's findings: first, the multitude of fixed stars (in particular those observed in the Milky Way); second, that Saturn represented three conjoined stars; third, the phases of Venus; fourth, the rough surface of the Moon; and fifth, the four satellites of Jupiter.

The reason that Bellarmino's first question concerned the reality of "a multitude of fixed stars" almost certainly had to do with disturbing memories related to the Giordano Bruno affair. Bruno's assertion that the universe is infinite and harbors a huge ensemble of inhabited worlds was one of the reasons that had led to his condemnation and tragic fate. Bellarmino had participated in the proceedings of that condemnation. Galileo's claimed discovery that the Milky Way was teeming with previously unseen stars undoubtedly gave Bellarmino a strong and unpleasant sense of déjà vu.

On March 24, 1611, Fathers Christopher Clavius, Giovanni Paolo Lembo, Odo van Maelcote, and Christoph Grienberger gave their answers: "It is true that with the spyglass very many wonderful stars appear in the nebulosities of Cancer and the Pleiades." The mathematicians were a bit more cautious about the Milky Way, acknowledging that "it cannot be denied that . . . there are many minute stars" but noting that "it appears more probable that there are continuous denser parts." As we know today, the Milky Way does contain in addition to the hundreds of billions of stars a disk of gas and dust. In the case of Saturn, the Jesuit mathematicians confirmed the oOo shape observed by Galileo, adding, "we have not seen the two starlets on either side sufficiently separated from the one in the middle to be able to say that they are distinct stars." They fully confirmed the waning and waxing phases of Venus and the fact that "four stars go about Jupiter, which move very swiftly." The only observation about which they had some reservations was that of the Moon. They wrote:

"The great inequality of the Moon cannot be denied. But it appears to Father Clavius more probable that the surface is not uneven, but rather that the lunar body is not of uniform density and has denser and rarer parts, as are the ordinary spots seen with the natural light. Others think that the surface is indeed uneven, but thus far we are not certain enough about this to confirm it indubitably."

The opinion of the Catholic Church's most prestigious mathematicians marked an incredible victory for Galileo. Clavius's thoughts on the interpretation of the lunar observations notwithstanding, the Collegio Romano scientists recognized the telescope as a bona fide scientific instrument that delivers reality in finer detail. One could no longer argue that the telescope deceives or presents a misleading picture of the cosmos. From then on, all serious discussion could concern only the interpretation and meaning of the results rather than the telescope itself or the reality of the discoveries made with it.

The current debate on global warming had to go (and to a large extent it is still going) through a similarly painful type of confirmation process. First, people have to be persuaded that the phenomenon itself is real; then they have to accept that the identification of its causes is correct; and finally, they have to embrace at least some of the recommended solutions.

As Galileo's case (and, indeed, those of Darwin, Einstein, and other scientists) have demonstrated, we should trust the science—the stakes are simply too high not to. We can, and should, have a serious discussion on precisely what to do in order to address the consequences of scientific discoveries, such as the threats posed by climate change (for example, rising sea levels and the dramatic increase in the frequency of extreme weather events). There should, however, be no debate anymore on whether climate change is real, on what is causing it, and on whether doing nothing is an option.

Ironically, some climate change deniers have even tried to argue that the overwhelming consensus in the geoscience community about a human-caused climate change is in itself a "logical fallacy," citing Galileo's case. The argument goes as follows: because Galileo was mocked and persecuted for his views by a majority at his time but later proved to be right, current minority views on climate change that are being criticized must also be right. In fact, this false logic even has a name: the "Galileo gambit." The flaw in the Galileo gambit is obvious: Galileo was right not because he had been mocked and criticized but because the *scientific evidence* was on his side. The reports on climate change by the leading scientific organizations represent, with obvious uncertainties which are clearly articulated, the up-to-date state of knowledge in this field. In science, we know that the near 100 percent consensus does not in itself guarantee that the conclusions are correct, but we know that this consensus is based on continually tested scientific evidence.

Returning to Galileo's case proper, what happened—for a while, at least—was that he was showered with honors. Overall, some 400 books about Galileo, approximately 40 percent of them favorable, were published in the seventeenth century alone. Of those, about 170 appeared outside of Italy. Even his archenemy Martin Horky, impressed by Galileo's observations of Saturn, declared his deep regret over having attacked such a magician of the heavens. In fact, he went so far as to say that he would have preferred to lose blood.

On April 14, 1611, during a banquet at which Galileo was elected as a sixth member of Cesi's Accademia dei Lincei, the name *telescopium* was coined for the spyglass that revolutionized cosmology. The name was suggested by theologian and mathematician Giovanni Demisiani, and it didn't take long for the first book on the history of the telescope to appear. It was written by the Milanese Girolamo Sirtori in 1612 and published in 1618 under the obvious title *Telescopium.*

Galileo's triumphant reception in 1611 represented just one battle. It did not mean that he had won the war. While the veracity of the observations themselves had been accepted, this merely marked the starting point for the altercation over the interpretation of the results. It was to be expected that staunch, fervent geocentrists, compelled to reevaluate cherished beliefs about the cosmos and the Earth's status within it, were not going to capitulate without a fight.

As we know today, the Copernican model as advocated by Galileo marked the introduction of a new concept, now known as the "Copernican principle": the realization that the Earth, and we, human beings, are nothing special, from a physical perspective, in the grand scheme of things. In the centuries that have passed since Copernicus's proposed scenario and Galileo's discoveries, this principle of cosmic humility has only gained strength through a series of steps demonstrating that, indeed, we do not occupy any special place in the cosmos.

First, Copernicus and Galileo removed the Earth from its central position in the solar system. Then, in 1918, astronomer Harlow Shapley showed that in the Milky Way Galaxy, the solar system itself isn't central at all. It is almost two-thirds of the way out; literally in the galaxy's remote suburbs. And in 1924 astronomer Edwin Hubble discovered that there are many other galaxies in the universe. In fact, we know today that there are perhaps as many as two trillion galaxies in the observable part, according to the latest astronomical estimate. As if this wasn't enough, some cosmologists now speculate that even our entire universe may be just one member of a huge ensemble of universes—a *multiverse.*

An interesting case illustrating what Galileo was up against in his endeavors to prove the superiority of the Copernican cosmol-

ogy over the Aristotelian (or Ptolemaic) one was provided by Cesare
Cremonini, a renowned philosopher and dogmatic Aristotelian.
Cremonini was Galileo's colleague at the University of Padua, where
the two often engaged in a form of friendly rivalry. He was extremely
opinionated when it came to natural philosophy, to the point that
he apparently was afraid to even put William Gilbert's book on
magnets and the magnetic Earth on his shelf, for fear that it would
contaminate his other books. While Cremonini was an atheist, and
also forcefully defiant of censorship, he felt obligated to defend Aristotelianism in all its forms. Consequently, he challenged Galileo's
assertion that the nova of 1604 was farther out than the lunar orbit—
because that went against Aristotle's doctrine that all changes in the
heavens were sublunar. When Galileo offered to show Cremonini
his new discoveries, Cremonini was said to have refused to even look
through the telescope (as did, by the way, the chief philosopher at
Pisa, Giulio Libri). This behavior won Cremonini the dubious honor
of Galileo patterning the stubborn Aristotelian Simplicio partly after
him in the *Dialogo*. In reality, Cremonini wanted something deeper
than what had been revealed by Galileo's observations. He noted,
for instance, that if the Moon was indeed a terrestrial-like body, as
Galileo's findings had implied, it should have fallen toward the Earth.
In the absence of a theory that could explain why this was not happening (a situation that lasted until Newton), Cremonini was not
prepared to give up his Aristotelian views.

Galileo was never a great believer in what he considered to be
unseen spooky forces, such as what Newton would eventually identify
as a gravitational force, acting at a distance. This fact played a role in
his later theory of ocean tides. Even when discussing Gilbert's experiments with magnetism, which involved a somewhat mysterious, un-
observed magnetic force, he wished for Gilbert to have come up with
an explanation "well grounded in geometry" for his findings, since he
found the reasons given by Gilbert "not compelling with the strength
which those adduced for natural, necessary, and eternal conclusions
should undoubtedly possess."

In short, Galileo was at that moment incapable of producing a genuine theory of gravity, and therefore he remained perpetually concerned that even if "something beautiful and true were discovered, it would be suppressed by their [the philosophers'] tyranny."

There was another concept, by the way, with which Galileo had difficulties. Kepler had discovered that a circular orbit failed to fit Tycho Brahe's very detailed observations of Mars, which had been collected for more than thirty-eight years. Consequently, he reluctantly altered his model for the orbit and made it an ellipse. To his surprise, he found that not only did an elliptical orbit explain the motion of Mars but also of the other planets. This turned out to be one of Kepler's major discoveries, and he described it in his book *Astronomia Nova*, which was published in 1609.

Galileo never accepted the idea of elliptical orbits. In this, even he—arguably the founder of modern scientific thought—remained prisoner of the ancient Platonic concept that perfect motion had to be circular. Today we know that it is not the *shape* of the orbit that needs to be symmetrical (not change) under rotations. Rather, it is the *law of gravity* that is symmetrical—meaning that the orbit can have any orientation in space.

The enormous efforts during those years that Galileo had put into the observations themselves, the books describing the findings, and the propaganda campaign for the dissemination of the discoveries, took a heavy toll both on his health and his family life. Being as driven as he was, he probably cared more about the former than about the latter. As a result of heavy drinking and unhealthy food and lifestyle, he suffered from various rheumatic pains, fever, and irregular heartbeats during the winter of 1610 and the summer of 1611. Not just Horky noticed his sickly, sallow appearance—a Venetian ambassador who had not seen him for several years was shocked when he met Galileo in 1615.

On the family side, Galileo left his companion Marina Gamba behind when he moved to Florence. She died in August 1612, which left Galileo in charge of their three children. He promptly solved part of that problem by putting his two daughters in the convent of San Matteo in Arcetri. The Sisters of San Matteo belonged to the order of Poor Clares and were indeed in a state of miserable poverty most of the time. There was nothing unusual about young women being placed in convents at that time. This was particularly true in the case of illegitimate daughters, whose prospects for marriage were limited, given that the sizeable dowry that would have been required to ensure an acceptable husband was beyond Galileo's means. Still, the choice of San Matteo of all places remains somewhat of a puzzle, given that particular convent's extreme penury and its location outside the city, which made it much harder to supervise the everyday behavior of men inside the convent's walls. Several cases of scandalous affairs between nuns and either unscrupulous father confessors or laymen who visited the convent have been known to occur. It is possible that the choice of the convent was imposed on Galileo by the fact that his daughters were really too young to be accepted as nuns. Galileo succeeded in getting them admitted only with the help of Cardinal Ottavio Bandini.

While we know very little about the life of Galileo's daughter Virginia (Sister Maria Celeste) until 1623, about 120 letters she wrote to Galileo between 1623 and 1634 have survived. From these, a picture of an extremely sensitive and caring daughter emerges. Being an apothecary at the convent, Maria Celeste used to send Galileo herbal treatments for his numerous maladies, and she even restocked his house with wine when he finally returned home after his trial by the Inquisition. Sadly, she died at age thirty-three from dysentery. The brokenhearted Galileo wrote about his daughter that she was "a woman of exquisite mind, singular goodness, and most tenderly attached to me."

Much less is known about Galileo's other daughter, Livia (Sister Arcangela), and even that, only from Sister Maria Celeste's letters to her father. It appears that Livia never adjusted to convent life, and her relationship with Galileo was seriously strained by the harsh conditions she experienced.

The fate of Galileo's son, Vincenzo, was much happier, primarily because the gender bias prevailing at the time ensured that there were no special financial obligations involved with a son. Vincenzo was eventually legitimized by the grand duke and, ironically, completed medical school at the University of Pisa—the program from which his father had dropped out. In case you wonder, there are no descendants of Galileo today. His last great-great-grandson, Cosimo Maria, died in 1779.

MORE ADO TO INTERPRET INTERPRETATIONS

In 1613 Benedetto Castelli, Galileo's former student, was appointed professor of mathematics at the University of Pisa. That December, as the Tuscan court implemented its customary annual move to Pisa, Castelli was invited several times to dine with the Medicis. This led to that famous breakfast at which Castelli was asked to explain the significance of Galileo's discoveries and the merits of the Copernican system. To understand the backdrop to that event, we should realize that, in some sense, Galileo's outreach campaign had been too successful. Having heard of his discoveries, various people began opposing his ideas on various grounds. In Florence, philosopher Lodovico delle Colombe challenged essentially every book Galileo had written up to that point. Between late 1610 and early 1611, he composed a treatise entitled *Contro il moto della Terra* (*Against the Motion of the Earth*), in which he listed numerous biblical quotations that supposedly showed that the Earth was motionless. He even went so far as to form a "league" hostile to Galileo. Scholars in Pisa were similarly lining up along ideologies unsympathetic to Galileo, with arguments for defending the Aristotelian system rapidly converging with reasons based on faith. Consequently, Castelli's dining with the grand ducal family occurred during these already fairly charged times, and, significantly, present at the breakfast was also the Pisan professor of philosophy Cosimo Boscaglia, an expert on Plato, whose views on Galileo were suspect at the very least.

The initial conversation was friendly enough, general, and quite benign. Nevertheless, the Grand Duchess Christina, a strictly pious woman, was already wondering whether Jupiter's satellites were real or just "illusions of the telescope." Boscaglia, asked for his opinion, replied that their reality "could not be denied." He did, however, whisper to Christina more privately that Galileo's Copernican interpretation was more problematic, since "the motion of the Earth had in it something of the incredible, and could not occur, especially because the Holy Scripture was obviously contrary to that view."

Following the meal, as Castelli was on his way out, he was recalled by Christina to her chambers, where he found, in addition to the duchess and the duke, a few other guests, including Don Antonio de' Medici (an admirer of Galileo) and Professor Boscaglia. For the next two hours, Christina grilled Castelli on what she regarded as discrepancies between the concept of a moving Earth and Holy Scripture. From her manner, however, Castelli's judgment was that she did this only to hear his replies. Boscaglia didn't say a word.

While the entire event seemed to have ended favorably, Galileo was still concerned that Castelli might be placed in similar situations again. That was why he wrote his long and detailed *Letter to Benedetto Castelli*, in which he outlined his ideas about the handling of apparent contradictions between biblical texts and scientific discoveries. Even though written more than four hundred years ago, this *Letter to Benedetto Castelli* and the subsequent expanded version—*Letter to the Grand Duchess Christina*—both written by a serious scientist who, having lived in seventeenth-century Italy, was also a "sincere believer" (in the words of Pope John Paul II)—remain remarkable documents on the relationship between science and Scripture. We shall return to this topic, which is still of great current interest, in chapter 17.

Galileo started his letter by praising Castelli for his success as a professor, adding, "What greater favor could you desire than to see their Highnesses taking pleasure in reasoning with you, raising questions, hearing their solutions, and finally resting satisfied with your replies?" He then explained that the incident caused him to think more generally about "the carrying of Holy Scripture into disputes about physical conclusions"—in particular the passage in Joshua about the Sun stopping in its course, which appeared to contradict "the mobility of the Earth and stability of the Sun." Galileo's opening statement about the usage of biblical texts powerfully sets the stage for his subsequent arguments:

"Holy Scripture could never lie or err . . . its decrees are of absolute truth" (emphasis added). Nevertheless, Galileo added, "some of its interpreters and expositors may sometimes err in various ways, one of which may be very serious and quite frequent, [that is,] when they would base themselves always on the literal meaning of the words. For in that way there would appear to be [in the Bible] not only various contradictions but even grave heresies and blasphemies, since [literally] it would be necessary to give to God feet and hands and eyes, and no less corporeal and human feelings, like wrath, regret, and hatred, or sometimes even forgetfulness of things gone by and ignorance of the future."

Galileo continued by insisting that for it to be understood by common, uneducated people, Scripture had to use a language that could be accessible. Consequently, he argued: "Physical effects placed before our eyes by sensible experience, or concluded by necessary demonstrations, should not in any circumstances be called in doubt by passages in Scripture that verbally have a different semblance." Especially, Galileo noted, since one cannot have a situation where two truths contradict each other. "Hence," he suggested, "apart from articles concerning salvation and the establishment of the Faith, against the solidity of which there is no danger that anyone may ever raise a more valid and efficacious doctrine, it would be the best counsel never to add more [articles of faith] without necessity." To which he added that (already mentioned) compelling, powerfully cogent reasoning that he did not believe that "the same God who has given us our senses, reason, and intelligence wished us to abandon their use."

Galileo then moved on to the particular passage in the book of Joshua where he showed, believe it or not, that a literal interpretation of the text coupled to the Aristotelian-Ptolemaic model would have resulted in *shortening* the day rather than lengthening it, as Joshua intended! The reason for this unexpected result had to do with the "mechanics" of Aristotle's vision of the heavens. In Aristotle's scenario, the Sun participated in two motions: one was its own "private" annual motion from west to east, and the second, a motion of the entire sphere of stars (together with the Sun) from east to west. Stopping the Sun's "private" motion (from west to east) would have clearly shortened the day, since the Sun would move even faster from east to west. Halting the Sun alone, while allowing the heavenly sphere to revolve, would have literally upset the entire celestial order. In contrast, in the Copernican cosmology, simply barring temporarily the Earth's spin around its axis would have produced the desired effect.

There is no question that with today's hindsight, Galileo's logic seems crystal clear and robustly persuasive. In this sense, he was an even more forward-looking theologian than Cardinal Roberto Bellarmino and other contemporary church officials. Even Pope John Paul II noted that Galileo "proved himself more perspicacious on this issue than his theologian adversaries." It is important to remember, however, that to a large extent, the objection to Copernicanism had much less to do with the actual cosmological model—the Church was not particularly interested in which planetary orbits astronomers preferred to use—and more with what some Catholics, and church officers in particular, regarded as an unwelcome intrusion of scientists into theology. Consequently, in spite of Galileo's conviction that he had not only adequately addressed all the issues raised by Christina but also demonstrated that the truth can be hidden behind the appearances, the *Letter to Benedetto Castelli* and the interpretation of that passage from Joshua were going to come back to haunt him.

If you think that the problem of literal interpretations of old texts of any sort is entirely a thing of the past, think again. In his famous *Essays*, the French writer Michel de Montaigne recognized already in the sixteenth century that "there is more ado to interpret interpretations than to interpret the things, and more books upon books than upon all other subjects; we do nothing but comment upon one another." As US Supreme Court decisions have proven time and again, even today, interpretations remain as critically important as they were at Galileo's time. For Galileo himself, interpretations were about to become almost a matter of life and death.

CHAPTER 6

Into a Minefield

One of the major goals of physics today is to formulate a theory, sometimes dubbed the Theory of Everything, that would elegantly unify all the fundamental forces of nature (gravity, electromagnetism, and the strong and weak nuclear interactions). Such a theory should also self-consistently combine our current best theory of gravity and the universe at large (Einstein's general relativity) with the theory of the subatomic world (quantum mechanics).

Through his demonstration that celestial bodies and their characteristics are really no different from the Earth and terrestrial attributes, Galileo took the first, insightful step toward such a unification. He showed that the Sun has at its outer layers features (sunspots) that resemble atmospheric phenomena on Earth; that Jupiter (and perhaps Saturn) has even more moons than the Earth; that Venus exhibits phases like the Moon; that the Moon's surface is covered with mountains and plains like those of the Earth; and that the Earth itself reflects sunlight onto the Moon, just as the Moon brightens the Earth's nighttime sky. Following these discoveries, one could no longer talk about separate, distinct "earthly" and "heavenly" qualities. Galileo proved that unlike Aristotle's vision of a sacrosanct, immutable celestial sphere, the heavens are just as prone to change as the

Earth—as demonstrated, for instance, by the appearance of novae and comets. About eight decades later, these concepts, together with the mathematization of physics, were precisely the ingredients that opened the door for Newton's all-embracing theory of universal gravitation.

All of Galileo's awe-inspiring revelations might have been accepted as constituting incredible scientific progress, were it not for the unfortunate fact that they contradicted the Aristotelian cosmology, which the Catholic Church had adopted as its orthodoxy centuries earlier. What's more, the Copernican system was bound to be at odds with a worldview that had placed humans at the very center of creation, not only physically but also as a purpose and focus for the universe's existence. The resistance to the Copernican downgrading of the Earth and its inhabitants would partly explain the later objections to Darwinism—the other theory demoting humans from uniqueness and making them rather a natural product of evolution.

All of this notwithstanding, however, the Church might have still accommodated (albeit with difficulty) a *hypothetical* system that would have made it easier for mathematicians to calculate orbits, positions, and appearances of planets and stars as long as such a system could be dismissed as not representing a true physical reality. The Copernican system could be accepted as a mere mathematical framework: a model invented so as to "save the appearances" of astronomical observations—that is, to fit the observed motions of the planets.

The crucial act that really brought about the Church's wrath was what Catholic officials regarded as an unacceptable, impudent invasion into the Church's exclusive provinces—theology and the interpretation of Scripture. Consequently, even as the opposition to Galileo's findings on purely astronomical and natural philosophy grounds was starting to abate, the antagonism on account of the theological issues was about to grow.

The stage for the theological debate that was going to play a decisive role in the drama that has become known as the Galileo affair had been set almost a century earlier with the Protestant Reformation. That was the point at which a schism had developed concerning authority in interpretating the Bible. Consequently, the notion that literal readings of Scripture were essential and unassailable was rapidly gaining acceptance among Catholic theologians. The Dominican Scholastic theologian Domingo Bañez, for instance, had expressed his views in 1584: "The Holy Spirit not only inspired all that is contained in the Scripture, He also dictated and suggested every word with which it was written." Another Dominican theologian, Melchior Cano, went even further when he declared in 1585: "Not only the words but even every comma has been supplied by the Holy Spirit." And who had the authority to interpret those words? The Catholic Church had already in its arsenal of resources an empyrean answer for that too. The Council of Trent, which had been held between 1545 and 1563 as an embodiment of the fight to counter the Protestant Reformation, issued on April 8, 1546, an unambiguous decree: "In the matters of faith and morals pertaining to the edification of Christian Doctrine, no one, relying on his own judgment and distorting the Sacred Scriptures according to his own conceptions, should dare to interpret them contrary to the sense which Holy Mother Church, to whom it belongs to judge their true sense and meaning, has held and does hold, or even contrary to the unanimous agreement of the Fathers." Given this authoritarian, uncompromising language, it was becoming clear that Galileo's rationalizations in his *Letter to Benedetto Castelli* could attract the censors' attention.

In some sense, Galileo's statements about the impropriety of using biblical texts literally in order to contradict observational findings came at the worst possible time, when the Church was extraordinarily sensitive to any attempt to undermine its authority in the interpretation of Scripture. A conflict therefore appeared almost inevitable. Unfortunately, as we shall see in chapter 16, even as late as 1945, Vatican authorities interdicted the publication of a book on Galileo commissioned by the Pontifical Academy of Sciences itself, because they regarded it as too "pro-Galileo" in describing the affair.

In any case, Galileo's situation in 1615 was going from bad to worse, when the Florentine Dominican Niccolo Lorini sent on February 7 what he called a "true copy" of the *Letter to Benedetto Castelli* to Cardinal Paolo Camillo Sfondrati, the prefect of the Sacred Congregation of the Index, to be examined. The Congregation of the Index was the body that was supposed to block the distribution of any printed material deemed contradictory to the Catholic faith. In fact, since the *Letter to Benedetto Castelli* was not printed, the Congregation of the Index was the wrong address for it. However, since the letter did nevertheless concern matters considered to be related to faith, the prefect forwarded Lorini's letter together with the *Letter to Benedetto Castelli* to the secretary of the Holy Office, who immediately asked for the opinion of a consultor. Probably aware that the letter, which Galileo wrote to Castelli rather hastily, could spell trouble, Galileo produced a slightly revised version, in which he more thoughtfully and cautiously presented the theological issues. He then sent the letter with an explanation to his friend the Florentine monsignor Piero Dini. Galileo asked Dini to show the letter to the Collegio Romano mathematician Christoph Grienberger and, if appropriate, also to Cardinal Bellarmino, pointing out that "Nicolaus Copernicus was a man not only Catholic, but religious and a canon, and he was called to Rome under Pope Leo X when, in the Lateran Council, the emendation of the calendar was dealt with, he being utilized as a very great astronomer."

There has been a fascinating recent story concerning the *Letter to Benedetto Castelli*. The original version was long thought lost, but in August 2018 it was discovered in the Royal Society's possession in London, where it apparently has been for at least 250 years, escaping the notice of historians. The rediscovery was made by Salvatore Ricciardo, a postdoctoral science historian at the University of Bergamo, in Italy, who browsed the London Royal Society's online catalogue for a different purpose. From the differences between the existing versions, we can see Galileo's attempts to moderate the tone of the original letter. For example, Galileo originally referred to certain propositions in the Bible as "false if one goes by the literal meaning of the words." He then crossed out the word *false* and replaced it with "look different from the truth." He also changed his reference to the Scriptures "concealing" its basic dogmas to the less harsh "veiling." The reason that the letter had been overlooked by Galileo scholars may have been that when it was catalogued in 1940, it was misdated to 21 December 1618, instead of 1613.

A few of Galileo's friends recognized relatively early the potential risks that were brewing, and they cautioned him to tread lightly. Federico Cesi, the founder of the Lincean Academy, encountered the theological obstacles right away: when he tried to publish *Letters on Sunspots*, he failed in several attempts to include in the publication references to biblical texts or to Galileo's claims that the Bible was in fact more consistent with Copernican views than with Ptolemaic views. For instance, the censors insisted on eliminating the statement in Galileo's second letter to Markus Welser (which was probably based on the answer Galileo had received from Cardinal Carlo Conti), where he said that the immutability of the heavens was "not only false but erroneous and repugnant to the truths of Sacred Scripture about which there could be no doubt." Realizing that there was no way to get such comments past the censors, Cesi removed all allusions to the Bible from the publication. Galileo, however, may have not taken sufficient account at that stage of the significance of the censors' intervention with respect to theology.

Whereas all of Galileo's well-wishers were advising him to keep a low profile on all theological issues, his opponents were starting to become increasingly vocal. In particular, a fiery, aggressive preacher named Tommaso Caccini became one of a clique that caused the most damage. This particular episode also started with Galileo's foe Lodovico delle Colombe, who a few years earlier had the dispute with Galileo over the nova of 1604, and who in 1611 wrote the dissertation *Against the Motion of the Earth*, in which, to Galileo's dismay, he had drawn Scripture into the conversation. Lodovico, his Dominican brother Raffaello, and a few other Florentine Dominicans (a group known scornfully among Galileo's friends as the "Colombi," meaning pigeons) also obtained a copy of the *Letter to Benedetto Castelli* and attacked Galileo on account of his Copernican views, objecting too, to Galileo's discussion of spots on the Sun. Unfortunately, the delle Colombe brothers had the ear of Florence's archbishop and, through him, also of Caccini. The preacher seemed to have turned the act of proving that Galileo and the Copernicans were heretics into his mission in life. Toward this disturbing "goal," he delivered a fervent sermon on December 21, 1614, from the pulpit at Florence's Santa Maria Novella, in which, citing again that overused and abused passage from the book of Joshua, Caccini asserted that the Copernican system, with its central, unmoving Sun, "was a heretical proposition." This incident in itself might have passed relatively unnoticed—Caccini was reprimanded by both his brother, who was the head of the Caccini house, and by other Dominican officials— were it not for the fact that Caccini also went to Rome on March 20, 1615, to testify before the Dominican Michelangelo Seghizzi, the commissary general of the Holy Office. In his deposition, among many other deleterious statements, Caccini said emphatically, "It is a widespread opinion that the above-mentioned Galilei holds these two propositions: the earth moves as a whole as well as with diurnal motion; the sun is motionless." He added that these propositions were "repugnant to the divine Scripture."

Even worse, being aware that Paolo Sarpi was on the Inquisition's watch list because of his role in a dispute between the Venetian Republic and the Pope a decade earlier, Caccini included a malevolent comment emphasizing Sarpi's friendship with Galileo. Similarly, he deliberately and viciously mentioned that Galileo was corresponding with contemporaries in Germany, knowing that this would raise the specter of Lutheranism and guilt by association.

Around the same time, Castelli, who himself was starting to feel the heat in Pisa, wrote to Galileo to express his concerns, noting in low spirits and frustration, "I am most displeased that the ignorance of some people has peaked so that, condemning science of which they are totally ignorant, they attribute [false] things to science they

are incapable of understanding." Sadly, a similar attitude still charac-
terizes a few of the current climate change deniers, and quite a few of
the responses to the COVID-19 pandemic.

The disdain for and enmity toward science we are experiencing
today is precisely the type of attitude Galileo was fighting against.
Through his attempts to separate science from the interpretation of
Scripture and his reading of the laws of nature from experimental
results rather than associating them with a certain "purpose," Galileo was one of the first to implicitly introduce the idea that science
compels us to take responsibility for our own destiny, as well as our
planet's.

Castelli, after acknowledging the gloomy reality that he and Galileo were about to face, added in his letter: "But patience, for these
impertinences are neither the first nor the last." In a letter dated
January 12, Cesi expressed precisely the same sentiments, referring to
the attackers on Copernicanism as "the enemies of knowledge." Cesi
also took the opportunity to reiterate his advice to Galileo to remain
inconspicuous. His strategy for dealing with the onslaught was to
recruit other mathematicians and to present the entire affair as an
assault on mathematicians, rather than to try to advocate the truth
inherent in Copernicanism.

In the meantime, the *Letter to Benedetto Castelli* was continuing to
cause problems. The consultor that the Holy Office employed came
back with relatively minor reservations, and even those addressed
only three of the statements made in the letter, adding that "for the
rest, though it sometimes uses improper words, it does not diverge
from the pathways of Catholic expression." Unfortunately, this
benign judgment only incentivized the Holy Office to dig deeper. To
that effect, it asked the inquisitor of Pisa to obtain the original letter
from Castelli himself.

While all of this turmoil was happening, Monsignor Dini was
busy trying to help Galileo in any way he could. He gave copies of
the slightly revised *Letter to Benedetto Castelli* to both Grienberger
and Cardinal Bellarmino, and he consulted about the entire situation

with the young church official and poet Giovanni Ciampoli, who knew Galileo and had been a childhood friend of Cosimo II de' Medici's. Ciampoli had been freshly ordained as a priest in Rome in 1614. In response to Dini's appeal, he transmitted directly to Galileo advice from Cardinal Maffeo Barberini (later to become Pope Urban VIII), saying that "he [Barberini] would like greater caution in not going beyond the arguments used by Ptolemy and Copernicus, and finally in not exceeding the bounds of physics and mathematics. For to explain the Scripture is claimed by theologians as their field, and if new things are brought in even though to be admired for their ingenuity, not everyone has the dispassionate faculty of taking them just as they are said." In other words, Cardinal Barberini's unequivocal recommendation was that Galileo should stay away from any new interpretations of the Bible.

Similar directives were coming from Cardinal Bellarmino, too, also via Dini. The cardinal's assessment was that Copernicus's book *On the Revolutions of the Heavenly Spheres* would not be prohibited, but that a note would be added, to present the Copernican system as merely a mathematical model. Bellarmino suggested further that Galileo should adopt the same stance, since, he noted, the biblical text in Psalms 19:5–6 stood, in his view, in clear contradiction to the notion of a Sun standing still: "In the heavens, he has pitched a tent for the sun, which is like a bridegroom coming forth from his pavilion, like a champion rejoicing to run his course. It rises at one end of the heavens and makes its circuit to the other, nothing is hidden from its heat." Dini himself protested that this text could be interpreted as a poetic way of speaking, but Bellarmino retorted that this was "not something to jump into, just as one ought not to jump hurriedly into condemning any one of these opinions."

Unconvinced, Galileo sent a long answer to Dini on March 23, 1615, in which he tried to address Bellarmino's comments. He started by pointing out that in the biblical description in Genesis, light was created before the Sun. He then suggested that this light "unites and fortifies itself in the solar body," which needs to be at the center of the universe, because it "diffuses this light and prolific heat that gives life to all the members that lie around it." With regard to the passage from Psalms, Galileo argued that the implied motion was of the radiation and caloric spirit, "which, leaving from the solar body, is swiftly diffused throughout the entire world," and not of the Sun itself. Finally, since he had no theory of gravitation, Galileo used his discovery of the Sun's rotation about its axis to suggest a rather far-fetched model (knowing what we know today), in which this rotation was somehow driving the revolutions of the planets around the Sun. Since by writing this letter Galileo basically ignored all the cautionary advice given to him, Dini wisely decided (after consulting with Cesi) not to give this response to Cardinal Bellarmino.

Let's, however, think for a moment about what Galileo's friends and all the Church officials who were not (at least not yet) unsympathetic to him were advising him to do. In Galileo's view, even though at that point he still lacked direct proof for the Earth's motion, his discoveries had already achieved two things: First, some of the arguments of those who had claimed they had proof that the Earth was *not* moving (such as, that the Earth would have lost its Moon) had largely been shown to be false. Second, Galileo's findings constituted for him so much of a "smoking gun" for the Copernican system that there was no question in his mind that this model had to be considered as being at least potentially correct. And correct not just as some mathematical abstraction that happened to somehow mimic nature, but as a true description of physical reality.

Galileo was fighting here opinions frozen by centuries in which science had been considered detached from observations. The term "to save the appearances" had been coined to describe scientific models that conveniently simplified observations but had no deeper significance. Bellarmino, Grienberger, Barberini, and others were asking Galileo to give up convictions that had been forged on the basis of painstaking scientific observations and brilliant deductions, only because they *appeared* to contradict some sacred, ancient, vague, poetic texts—and only when those texts were interpreted literally rather than figuratively. In other words, it is not true that Bellarmino and Grienberger were trying only to convince Galileo not to meddle in theology, as a few modern scholars have concluded. This is demonstrated, for instance, by the fact that when addressing the arguments that Galileo had presented in favor of Copernicanism, Grienberger told Dini that he was "worried about other passages of the Holy Writ," and Bellarmino specifically mentioned that the Copernican doctrine should be presented solely as a pure mathematical expedient. Far from being annoyed merely by Galileo playing the theologian and his foray into biblical exegesis, these individuals were quite intent on crushing the Copernican challenge as a representation of reality because, from their perspective, they were vindicating the authority of Scripture in determining truth.

Can one be surprised, then, that Galileo refused to cooperate, at least initially? Should he have abandoned what he regarded as the only possible logical conclusions in favor of what amounted to a seventeenth-century version of political correctness? Recall that Galileo was right after all. Galileo never raised any doubts about the veracity of the biblical texts. At that point, he was still hoping for reason to prevail, and he did his best to prove that while interpretations of Scripture could be reformulated in alternative ways so as to agree with what nature was presenting, facts were facts.

UNEXPECTED SUPPORT WITH UNEXPECTED CONSEQUENCES

Galileo's more open support for Copernicanism, starting in 1615, which went against the judgment of his friends and the advice he had received from church officers, was likely motivated, influenced, and encouraged by a surprising booklet published by Carmelite theologian Paolo Antonio Foscarini.

A native of Montalto Uffugo in Calabria, Foscarini had been known to have broad knowledge in topics ranging from theology to mathematics. Cesi sent Galileo a copy of Foscarini's book on March

7, 1615. The very short publication had a very long title, part of which read *Letter of the Reverend Father Master Antonio Foscarini, Carmelite, on the Opinion of the Pythagoreans and of Copernicus Concerning the Mobility of the Earth and the Stability of the Sun and the New Pythagorean System of the World, etc.* The title referred to the fact that the first nongeocentric model of the cosmos was indeed suggested by the followers of Pythagoras—Pythagoreans—in the fourth century BCE. The philosopher Philolaus proposed that the Earth, Sun, and planets all moved in circular orbits around a central fire. Greek philosopher Heraclides of Pontus added, also in the fourth century BCE, that the Earth rotated around its axis too, while Aristarchus of Samos was the first to propose a heliocentric model in the third century BCE.

In terms of its logical exposition, Foscarini's book was outstanding. He explained that there was no doubt that Galileo's discoveries made the Copernican system demonstrably much more plausible than the Ptolemaic one. Assuming that Copernican cosmology was correct, and taking for granted that Scripture always represented the truth, Foscarini argued that there clearly could be no conflict between them, since there is only one truth. He concluded therefore that it had to be possible to reconcile those seemingly problematic biblical passages with Copernicanism. This was precisely what Galileo had been claiming all along. Foscarini, by examining many of the contentious biblical paragraphs and grouping them into six categories, was able to offer specific exegetical principles that he thought could be used to eliminate all the apparent contradictions. Foscarini's motivation for writing the book was also remarkable: if Copernicanism were proven to be correct in the future, he argued, the Church would be able to use his new interpretations of the controversial texts to escape the inadmissible verdict that the Bible is wrong.

In conclusion, Foscarini made two important observations. First, regarding the interpretation of the biblical language:

Scripture serves us by speaking in the vulgar and common manner; for from our point of view, it does seem that the earth stands firmly in the center and that the sun revolves around it, rather than the contrary. The same thing happens when people are carried in a small boat on the sea near the shore; to them it seems that the shore moves and is carried backwards, rather than that they move forwards, which is the truth.

Foscarini's second significant point was quite astonishing in its boldness: "The Church," he said, "cannot err, in matters of faith and our salvation only. But the Church can err in practical judgments, in philosophical speculations, and in other doctrines which do not involve or pertain to salvation."

Cesi thought that Foscarini's book "could not have appeared at a better time, unless to increase the fury of our adversaries is damaging, which I do not believe." Galileo's subsequent actions indicate that he believed the same, at least initially. Unfortunately, they were both wrong. Church official Giovanni Ciampoli, later the correspondence secretary of Pope Urban VIII and a member of the Lincean Academy, predicted in a letter he wrote to Galileo on March 21, 1615, that Foscarini's book would be condemned by the Holy Office. (Ciampoli may have had some inside information.)

The first reaction to Foscarini's book came in the form of the opinion of an unnamed theologian. In the first paragraph, he labeled Foscarini's views on Copernicanism as "rash." In his documented *Defense*, Foscarini vehemently rejected this characterization, stating forcefully again that there is a clear distinction between matters of faith and morals, and those related to natural philosophy and science. Concerning the latter, Foscarini repeated his position that "the Sacred Scriptures ought not to be interpreted otherwise than according to what human reason itself established from natural experience and according to what is clear from innumerable data."

Foscarini sent a copy of the book and of his *Defense* to Cardinal Bellarmino for comment, and Bellarmino answered on April 12, 1615, emphasizing three points:

First, it seems to me that Your Paternity and Mr. Galileo are pro-
ceeding prudently by limiting yourselves to speaking suppositionally and not absolutely, as I have always believed that Copernicus
spoke. For there is no danger in saying that by assuming the earth
moves and the sun stands [the Copernican model], one saves all
the appearances better than by postulating eccentrics and epicycles
[the Ptolemaic model]; and that is sufficient for mathematicians.

This language clearly implied more a form of advice rather than
praise, to both Foscarini and Galileo—even though Bellarmino's letter wasn't even addressed to Galileo.

"However," the cardinal was quick to add, "it is different to want
to affirm that in *reality* [emphasis added] the sun is at the center of
the world and only turns on itself without moving from east to west,
and the earth is in the third heaven [meaning the third orbit in terms
of its distance from the Sun] and revolves with great speed around
the sun." Bellarmino then explained why, in his view, claiming that
the Copernican scenario represented reality was "a very dangerous
thing." This was, he said, since it was "likely not only to irritate all
scholastic philosophers and theologians, but also to harm the Holy
Faith by rendering Holy Scripture false."

Bellarmino's second point had to do with interpretations of the
biblical texts. Here he started with something he regarded as obvious:
"As you know, the Council [Trent] prohibits interpreting Scripture
against the common consensus of the Holy Fathers." Then, however,
he delivered an exegetical bombshell. In response to Foscarini's claim
that the Holy Fathers' authority in interpreting the Bible applies only
to matters of faith and morals but not to topics such as the motion of
the Earth, Bellarmino offered a startling expansion of what should be
called "matters of faith":

Nor can one reply that this [the motion of the Sun or the Earth]
is not a matter of faith because of the subject matter, *it is still*

a matter of faith because of the speaker [emphasis added]. Thus, anyone who would say that Abraham did not have two sons and Jacob twelve would be just as much a heretic as someone who would say that Christ was not born of a virgin, for the Holy Spirit has said both of these things through the mouth of the Prophets and the Apostles.

Simply put, Bellarmino contended that not only is everything said in Scripture true, but that *everything*, including the most banal factual detail (as long as its meaning is clear) is also a "matter of faith"! Clearly, under this much broader definition of "matters of faith" by the most influential cardinal of the day, even the Earth's motion became a matter of faith.

Thirdly, Bellarmino admitted that "if there were a true demonstration that the sun is at the center of the world and the earth in the third heaven, and that the sun does not circle the earth but the earth circles the sun, then one would have to proceed with great care in explaining the Scriptures that appear contrary, and say rather that we do not understand them, than that what is demonstrated is false." However, Bellarmino also proclaimed: "But I will not believe that there is such a demonstration until it is shown to me," emphasizing that it would definitely not be sufficient "to demonstrate that by assuming the sun to be at the center and the earth in the heaven one can save the appearances." To add further weight to this last statement, the cardinal went on to say that it was King Solomon "who not only spoke inspired by God, but was a man above all others wise and learned in the human sciences" who wrote in Ecclesiastes 1:5: "The sun also ariseth, and the sun goeth down, and hasteth to his place where he arose." Consequently, Bellarmino concluded, it was highly unlikely that the Sun, in fact, did not move, especially since every scientist "experiences that the earth stands still" and sees "that the sun moves."

Bellarmino's answer to Foscarini has been scrutinized, analyzed, and interpreted by numerous Galileo scholars, and opinions on it span the entire range from high praise, claiming that Bellarmino had demonstrated the open-mindedness of a forward-thinking scientist who anticipated the relativism of later centuries, to complete dismissal, arguing that he exhibited conservative narrow-mindedness. We shall return to the theological points later, but for the moment, let's concentrate more critically on Bellarmino's scientific reasoning.

His opening statement appeared to be quite promising: "if there were a true demonstration that the sun is at the center . . . then one would have to proceed with great care in explaining the Scriptures." In fact, had he finished with this passage, he would have exhibited an intuitive awareness of what was to become a guiding principle in science: when new observations contradict existing theories, the theories need to be reexamined. The problem was that he immediately followed that paragraph with text indicating that he believed such a demonstration to be eternally unachievable. Bellarmino gave a few reasons for this misguided conviction, all patently nonscientific. First, he stated that "saving the appearances" in astronomy did not constitute proof of the Earth's motion. Even this seemingly convincing point went against genuine scientific thinking. If two different theories explain *all* the observed facts equally well, scientists would prefer to adopt, even if tentatively, the simpler one. Following Galileo's discoveries, such a process would have definitely favored the Copernican system over the Ptolemaic one, which was what Galileo had been championing all along. The requirement of simplicity would have also given an advantage to Copernicanism over Tycho Brahe's hybrid geocentric-heliocentric model. Of course, the ultimate test would have been to find direct proof for the Earth's motion or for the two theories to make some predictions that could then be tested by subsequent observations or experiments. In contrast, Bellarmino preferred to stick with the theory favored by the Church's orthodoxy.

The cardinal's second argument had really nothing to do with science. It advocated a blind acceptance of authority: on one hand, through the adoption of the interpretation of the Holy Fathers, and on the other, relying on the presumed infinite wisdom of King Solomon, who supposedly had written the book of Ecclesiastes. Both of these reasonings were manifestations of an attitude that was completely foreign to the spirit of science and entirely antithetical to what Galileo was espousing. In other words, far from being a forward-thinking scientist, faith trumped science in Bellarmino's world.

Finally, Bellarmino's third remark represented a misunderstanding coupled with parochial thinking. He declared that we all experience that the Earth does not move, rather than recognizing that all we can say is that it *appears* to us not to move. To prove his point, and referring to the example given by Foscarini in his book, he claimed that "when someone moves away from the shore, although it appears to him that the shore is moving away from him, nevertheless he knows that this is an error and corrects it, seeing clearly that the ship moves and not the shore."

Following ideas he inherited from Copernicus, Galileo could not accept this line of reasoning. In the same way that one couldn't tell whether it was the Sun or the Earth that was moving, only that there was relative motion between them, he insisted that no experiment performed inside a sealed room moving at a constant speed along a straight line could tell whether you are standing still or moving. This insight is familiar to anybody who watches from a train window another train moving on a parallel track. It later became an essential pillar of Einstein's theory of special relativity, in which he showed that the laws of physics are the same for all observers moving at a constant relative velocity. One can, of course, argue that Bellarmino could not have anticipated in the seventeenth century what Einstein would discover and prove centuries later, but Bellarmino's position was extremely rigid. He did not believe that a proof of Copernicanism *could ever be found*. This was in stark contrast to what even Foscarini, a theologian himself, spelled out perspicuously: "Since something new is always being added to the human sciences, and since many things are seen with the passage of time to be false which previously were thought to be true, it could happen that, when the falsity of a philosophical opinion has been detected, the authority of the Scriptures would be destroyed." That is to say, while Foscarini understood that new discoveries and the knowledge gained from science could render models prevailing at the time (and thereby biblical interpretations) false, Bellarmino hid behind a dogmatic assurance.

Galileo addressed some of the theological issues in detail in his *Letter to the Grand Duchess Christina*, but it is worth pointing out that Bellarmino's letter to Foscarini contained ab initio a surprisingly weak argument even when it came to theology. This forced him to adopt something that we would refer to today as the "nuclear option." Bellarmino relied on the "common consensus of the Holy Fathers" and on the decree by the Council of Trent. However, as both Galileo and Foscarini noted perceptively, the council's pronouncement had specifically spoken about "matters of faith and morals, pertaining to the edification of Christian Doctrine," while the Earth's motion had nothing to do with faith or morals, nor had the Holy Fathers ever discussed or reached consensus on this topic. Apparently, even Bellarmino himself was aware of this shortcoming of his reasoning, since otherwise there is no convincing explanation for his amplification of the definition of "matters of faith" far beyond the usual religious issues, to include essentially everything in the Bible.

Galileo managed to see Bellarmino's letter to Foscarini, and at one point, he even articulated a response in a series of notes that he may have intended to send to Foscarini. But those undated notes were never published. Galileo's main point addressed Bellarmino's new, sweeping definition of "matters of faith" with razor-sharp logic:

> It is replied that then everything which is in Scripture is a "matter of faith because of who said it," and thus in this respect ought to be included in the regulations of the Council [of Trent]. But this is clearly not the case, because then the Council ought to have said, "The interpretations of the Fathers must be followed for every word in the Scriptures," rather than "in matters of faith and morals." Thus having said "in matters of faith," it seems that the Council's intention was to mean "in matters of faith because of

the subject matter." It would be much more "a matter of faith" to
hold that Abraham had sons, and that Tobias had a dog, because
the Scriptures say so, than to hold that the earth does not move,
granting that the latter is found in the Scriptures themselves. The
reason why the denial of the former, but not of the latter, would
be a heresy is the following. Since there are always men in the
world who have two, four, six, or even no sons, and likewise since
someone might or might not have dogs, it would be equally credible that someone has sons or dogs and that someone else does
not. Hence there would be no reason or cause for the Holy Spirit
to state in such propositions anything other than the truth, since
the affirmative and the negative would be equally credible to all
men. But this is not the case concerning the mobility of the earth
and the stability of the sun, which are propositions far removed
from the apprehension of the common man. As a result, it has
pleased the Holy Spirit to accommodate the words of the Sacred
Scripture to the capacities of the common man in such matters
which do not concern his salvation, even though in nature the
fact be otherwise.

Put simply, Galileo argued—albeit only privately—that the meaning
in the case of Abraham's sons or Tobias's dog is obviously literal, and
therefore believing it (or not) could be taken to be a matter of faith,
while the stability of the Earth is only figurative and hence not a
matter of faith. There is little doubt that Galileo chose the example of
Tobias's insignificant dog as something that is totally inconsequential
in religious matters.

Galileo would later answer attempts to prohibit any revisions to
interpretations of Scripture with a quote from renowned ecclesiastical historian Cardinal Cesare Baronio, who died in 1606: "The intention of the Holy Spirit is to teach us how one goes to heaven and not
how heaven goes." With a not so subtle jab at Bellarmino's letter to
Foscarini, Galileo added that he had doubts "whether it is true that
the Church obliges one to hold as articles of faith such conclusions

about natural phenomena" and that he believed that "it may be that those who think in this manner *may want to amplify the decree of the Councils in favor of their own opinion*" [emphasis added]. Indeed, a few of Bellarmino's eventual actions, or, rather, lack thereof, when the decree of the Congregation of the Index against Copernicanism was published on March 5, 1616, showed that he was in agreement with that decision.

This tumult was not encouraging. In spite of Foscarini's honest motives and thoughtful arguments, his book attracted more scrutiny to the Copernican issue, and that, combined with the damaging acts of Caccini, delle Colombe, and Lorini, generated an atmosphere in which the specter of a condemnation of Copernicanism by the Church was rapidly becoming a reality. To counter that disconcerting trend, Galileo composed his *Letter to the Grand Duchess Christina*, which was a powerful document defending the autonomy of scientific research. However, probably realizing the danger of his situation, Galileo prudently refrained from circulating yet another polemic. Instead, he decided to go to Rome to present his case in person, contrary to the advice of his friends there, all of whom suggested to postpone such a visit and "to be quiet." The Tuscan ambassador to Rome, Piero Guicciardini, was especially displeased with Galileo's plan, noting that "this is no fit place to argue about the Moon, or, especially in these times, to try to bring in new ideas." It goes without saying that this attempt at dissuading Galileo, who always believed in his power of persuasion, also fell on deaf ears, and he arrived in Rome on December 11, 1615.

CHAPTER 7

This Proposition Is
Foolish and Absurd

In Rome, Galileo was starting to realize the magnitude of the opposition he was facing. It was rapidly becoming apparent that what was badly needed was a clear demonstration or proof of the Earth's motion. Sensing that, Galileo formulated in January 1616 a theory of ocean tides, which may have been based on earlier ideas of his friend Paolo Sarpi. He outlined the theory in a letter entitled *Discourse on the Tides*, which he sent on January 8 to a very young cardinal, Alessandro Orsini, who was to become Galileo's supporter.

Galileo's theory of the tides must have sprung, at some level at least, from his or Sarpi's observations of water sloshing back and forth at the bottom of a barge on his trips from Padua to Venice. He noticed that when the barge was accelerating, water piled up at the back, and when it slowed down, it accumulated at the front. This to-and-fro motion, Galileo thought, resembled the tides. It then occurred to him that in the case of the Earth, the speeding up could be the result of the diurnal spin motion being in the same direction and combining with the velocity of the Earth in its revolution around the Sun, which happens once a day at a given point on the Earth's surface, as in point A in Figure 7.1. The slowing down, in this picture, occurs (again, once a day) when the velocities associated with the orbital motion and with the spin are in opposite directions (as in point B in Figure 7.1). The continents were assumed not to be dislocated by the combinations of these two motions, but the oceans were supposed to respond by sloshing. Galileo was therefore convinced that in the absence of even one of these two motions, "the ebb and flow of the oceans could not occur."

Unfortunately, in spite of the fact that Galileo thought that he had elegantly associated the Earth's motion with the tides, "taking the former as the cause of the latter, and the latter as a sign of and an argument for the former," his theory of the tides was neither correct nor convincing. The Earth's revolution around the Sun played a rather subordinate role in it, and it certainly couldn't explain actual observations of tides in the Adriatic Sea, where local conditions and secondary causes produced significant effects. The theory did conform to

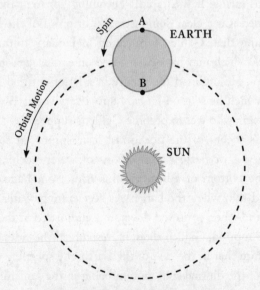

Figure 7.1. Schematic demonstrating
Galileo's theory of tides.

Galileo's general tendency to exclude the action of unseen forces acting across large distances, such as the gravitational attraction of the Moon, even though ideas along those lines had existed since antiquity, and the Flemish mathematician Simon Stevin, as well as Kepler, specifically suggested the Moon's attraction as the cause of the tides in 1608 and 1609, respectively. Albeit wrong, Galileo's commitment to mechanically easy-to-understand causation made his theory of tides at least plausible. Newton eventually used his theory of gravitation to explain in detail how the combined action of the gravity of the Moon and the Sun provides the tide-generating forces.

In an attempt to persuade a few of his adversaries, Galileo met with Caccini at the beginning of February 1616, but he did not succeed in appeasing him or in convincing him to change his views. He also discovered a new opponent, Monsignor Francesco Ingoli, who in January 1616 wrote an essay entitled "Disputation concerning the Location and rest of Earth against the system of Copernicus," and who was to become an active anti-Copernican.

Things took a turn for the worse when on February 19 the theologian consultors to the Holy Office were asked to give their opinion on two propositions: (1) The Sun is the center of the world and completely devoid of local motion, and (2) the Earth is not the center of the world, and not motionless—it moves as a whole of itself, and also with a diurnal rotational motion. Ironically, the same office that had objected vehemently to scientists intruding into theology was now asking the theologians to judge on two purely scientific questions—two of the central tenets of the Copernican model.

The consultors included the archbishop of Armagh, Ireland, the master of the Sacred Apostolic Palace, the Commissary of the Holy Office, and eight other religious figures, most of them Dominicans. Not one was a professional astronomer or even an accomplished scientist in any discipline. They took only four days to give their collective opinion. On the Sun being at the center of the solar system and motionless, they concluded that "this proposition is foolish and absurd in philosophy, and formally heretical, since it explicitly contradicts in many places the sense of Holy Scripture." They were slightly less harsh and more tentative on the second proposition, since the Bible doesn't say explicitly that the Earth does not move. Therefore, they concluded that "the proposition receives the same judgment in philosophy and that in regard to the theological truth, it is at least erroneous in faith." That is, they replaced the categorical "formally heretical" with "at least erroneous in faith."

Events then followed in rapid succession. Pope Paul V met with his cardinals on February 24. The freshly appointed Cardinal Alessandro Orsini, who was related to the Medicis through his mother, tried to argue in Galileo's defense and to present the scientist's theory of the tides. Orsini was greatly impressed with Galileo's arguments in a long conversation the two held about two months earlier. Unfortunately, he was cut short and promptly instructed by the Pope to convince Galileo to abandon his Copernican views. On the twenty-fifth, the Pope ordered Cardinal Bellarmino to summon Galileo and to warn him to renounce the opinion that the Sun stands still and the Earth is in motion. He added that refusal to obey the order would result in imprisonment. Bellarmino and Galileo met on February 26 in Bellarmino's chambers, in the presence of Michelangelo Seghizzi, the Commissary General of the Holy Office, and two other church functionaries from the cardinal's household. A document recorded by a clerk, summarizing what transpired at that meeting, became the centerpiece of the evidence at Galileo's trial seventeen years later:

At the palace of the usual residence of the said Most Illustrious Lord Cardinal Bellarmino and in the chambers of His Most Illustrious Lordship, and fully in the presence of the Reverend Father Michelangelo Seghizzi of Lodi, O. P. and Commissary General of the Holy Office, having summoned the above-mentioned Galileo before himself, the same Most Illustrious Lord Cardinal warned Galileo that the above-mentioned opinion was erroneous and that he should abandon it; and thereafter, indeed immediately, before me and witnesses, the Most Illustrious Lord Cardinal himself being also present still, the aforesaid Father Commissary, in the name of His Holiness the Pope and the whole Congregation of the Holy Office, ordered and enjoined the said Galileo, who was himself still present, to abandon completely the above-mentioned opinion that the sun stands still at the center of the world and the earth moves, and henceforth not to hold, teach, or defend it in any way whatever, either orally or in writing; otherwise the Holy Office would start proceedings against him. The same Galileo acquiesced in this injunction and promised to obey.

A second document describing what took place comes from the minutes of the meeting of the Holy Office on March 3. The account there reads: "The Most Illustrious Lord Cardinal Bellarmino having given the report that the mathematician Galileo Galilei had acquiesced when warned of the order of the Holy Congregation to abandon the opinion which he held till then, to the effect that the sun stands still at the center of the spheres but the earth is in motion."

The fact that the two documents, written on separate dates, contained some small but significant differences has generated much speculation among Galileo scholars. In particular, the meaning of the phrase "and thereafter, indeed immediately" in the first document is unclear. Was Galileo allowed to react to Bellarmino's initial admonition? If not, then there were no grounds for the Commissary General's injunction. If, as the second account seems to imply, Galileo promised to obey already after Bellarmino's warning, then, again, there was no reason for Seghizzi to intervene and impose a much more severe injunction (including not to "teach, or defend in any way whatever"). If one adopts a less conspiratorial interpretation, then the impression one gets is that maybe upon hearing Bellarmino's unexpected warning, Galileo hesitated a bit in his initial reaction, and that this precipitated an unjustified intervention by the impatient commissary general, who presented the injunction in more uncompromising terms. At that point, Galileo had to yield, or be imprisoned.

The Congregation of the Index also had to reach a decision as to what actions should be taken concerning publications related to the Copernican doctrine. The issue was again presented by Bellarmino, at meetings that took place at the beginning of March 1616. On March 5 the Congregation published its devastating decree:

> This Holy Congregation has also learned about the spreading and acceptance by many of the false Pythagorean doctrine, altogether contrary to the Holy Scripture, that the earth moves and the sun is motionless, which is also taught by Nicolaus Copernicus's *On the Revolutions of the Heavenly Spheres* and by Diego de Zúñiga's *On Job*. [The latter, a commentary by the sixteenth-century Augustinian hermit, concluded that the Copernican system was in better agreement with the book of Job than the Ptolemaic one, and that "the mobility of the Earth is not against the Scripture."] This may be seen from a certain letter published by a certain Carmelite Father, whose title is *Letter of the Reverend Father Master Paolo Antonio Foscarini, Carmelite, on the Opinion of the Pythagoreans and of Copernicus Concerning the Mobility of the Earth and the Stability of the Sun and the New Pythagorean System of the World, etc.*, in which the said Father tried to show that the above-mentioned doctrine of the sun's rest at the center of the world and the earth's motion is consonant with the truth and does not contradict Holy Scripture. Therefore, in order that this opinion may not creep any further to the prejudice of Catholic truth, the Congregation has decided that the books of Nicolaus Copernicus (*On the Revolutions of Spheres*) and Diego de Zúñiga (*On Job*) be suspended until corrected; but that the book of the Carmelite Father Paolo Antonio Foscarini be completely prohibited and condemned; and that all other books which teach the same be likewise prohibited, according to whether with the present decree it prohibits and condemns, and suspends them respectively.

At some level, the "good" news, from Galileo's perspective, was that he had not been mentioned by name, nor had his publications been criticized in the decree. Nevertheless, just one day before the decree's publication, the Tuscan ambassador, Guicciardini—who had previously advised against Galileo's visit to Rome—sent a letter to the grand duke with a strong "I told you so" tone: "He [Galileo] is all afire on his opinions and puts great passion in them, and not enough strength and prudence in controlling them; so that the Roman climate is getting very dangerous for him, and especially in this century, for the present Pope, who abhors the liberal arts and this kind of mind, cannot stand these novelties and subtleties; and everyone here tries to adjust his mind and his nature to that of the ruler." Simply put, Galileo received his first serious warning at the time of the strongly anti-intellectual Pope Paul V.

One can hardly miss the similarity between Guicciardini's description of the pervading mood in Rome of 1616 and that of today, replacing the word *Pope* with the appropriate current *ruler* who "abhors the liberal arts" and who "cannot stand these novelties and subtleties." This raises the critical question of whether freedom of thought, and decision-making based on informed, evidence-based reasoning, are sufficiently strong at present, so as to prevent both catastrophic consequences and modern versions of the Galileo affair from reoccurring. Unfortunately, history has shown that the practice of denying science because of one's beliefs has been repeated many times, even in the secular world.

Galileo was trying to make the best of a horrible situation by pointing out in a letter to the secretary of state to the grand duke that he believed that the changes to Copernicus's book would be minimal. In reality, the modifications introduced by Cardinal Luigi Caetani and later by Cardinal Francesco Ingoli were indeed relatively minor, and the publication of the revised version was approved in 1620. However, the new edition never reached the press, and so Copernicus's book remained on the *Index of Prohibited Books* until 1835! Nevertheless, Galileo was apparently correct in his assessment that, relatively speaking, the decree would not affect him too adversely, at least initially. In fact, he received an audience with the Pope just a week after the decree's publication, and the pontiff promised him that he could feel safe for as long as the Pope lived. Even more important, in the wake of circulating rumors that the Church had imposed on Galileo severe atonements, self-abasement, and abjuration of his Copernican ideas, Cardinal Bellarmino issued a quite remarkable letter on May 26, 1616, in which he affirmed the following:

> We, Roberto Cardinal Bellarmino, have heard that Mr. Galileo Galilei is being slandered or alleged to have abjured in our hands and also to have been given salutary penances for this. Having been sought about the truth of the matter, we say that the above-mentioned Galileo has not abjured in our hands, or in the hands of others here in Rome, or anywhere else that we know, any opinion or doctrine of his; nor has he received any penances, salutary or otherwise. On the contrary, he has only been notified of the declaration made by the Holy Father and published by the Sacred Congregation of the Index, whose content is that the doctrine attributed to Copernicus (that the earth moves around the sun and the sun stands at the center of the world without moving east to west) is contrary to Holy Scripture and therefore cannot be defended or held.

Obviously, Galileo was pleased with this document, and seventeen years later, he relied heavily on it for his defense when he was put on trial by the Inquisition. Nonetheless, we shouldn't get carried away appreciating Bellarmino for writing this favorable letter. While the cardinal was certainly not the person who decided the Church's view of Copernicanism, the fact remains that he did not object to the decree. Moreover, in spite of his seemingly moderate tone in his response to Foscarini, he did not argue (or at least did not do enough to convince the Congregation) to delay or postpone the decree until more observational evidence could be gathered, to avoid premature judgment. The net result of this lack of action on the part of Bellarmino and of all the mathematicians of the Collegio Romano—who had confirmed all of Galileo's findings—was an erroneous, ill-considered decision. The ruling was made by officers of the Church for whom retaining authoritative power over areas totally outside their expertise took priority over open-minded critical thinking informed by scientific evidence. Sadly, we don't lack modern-day equivalents.

Why did the Jesuit mathematicians remain silent? We shall probably never know for sure, but their passive attitude may have reflected a misguided form of scientific caution. There is no doubt that the Jesuit astronomers realized, as Clavius himself had admitted, that the Aristotelian doctrine was no longer tenable. In the absence of direct, convincing proof for the Earth's motion, however, the Jesuits may have opted to sit on the fence regarding the scientific question, relying on the fact that a compromise theory (Tycho Brahe's geocentric-heliocentric model) had not yet been ruled out definitively, and it was not in conflict with Scripture. In theological matters per se, the Jesuits could not compete with or claim superiority over the Dominicans. Be that as it may, the outcome was deplorable, and the situation was going to become even gloomier and more tragic with Galileo's trial in 1633. The fact remains that even at lectures starting the academic year at the Collegio Romano in 1623, Jesuit professors still spoke against "finders of novelties in the Sciences."

Over the past four centuries, there have been several attempts, especially in Catholic apologetic writings, to argue that some of the responsibility for the prohibition of Copernicanism lies with Galileo himself, because he wouldn't keep his mouth shut. Such claims are outrageous. As shown clearly in his *Letter to Benedetto Castelli*, his letters to Cardinal Dini, and his *Letter to the Grand Duchess Christina*, Galileo intended for the Church authorities to recognize Copernicanism—for which he saw compelling scientific evidence—as a potentially viable theory, and to defer judgment on it rather than to authoritatively and definitively condemn it. In his *Letter to the Grand Duchess Christina*, Galileo reaffirmed his belief in the truth of Scripture but emphasized the importance of interpretation: "Holy Scripture can never propose an untruth, always on condition that one penetrates to its true meaning, which—I think nobody can deny—is often hidden and very different from that which the simple signification of the words seems to indicate." Even if his declared religiousness was partly for tactical purposes, simply to defend himself, the logic of Galileo's argument must be admitted. Moreover, independently of Galileo, Foscarini had a similar goal, yet Ciampoli predicted correctly that Foscarini's book would be condemned.

The key point remains that, unlike in the history of art or even in the history of religious ideas, in the history of science, we can eventually know who was right. Galileo was right, and the Church in this case abused its disciplinary power. As Pope John Paul II admitted in 1992: "This led them [the theologians who condemned Galileo] unduly to transpose into the realm of the doctrine of the faith, a question which in fact pertained to scientific investigation." Such acknowledgments, however, didn't come for almost four centuries. In 1619 Galileo's already complex relationship with the Jesuit astronomers was about to seriously deteriorate.

CHAPTER 8

A Battle of Pseudonyms

Comets had fascinated people since antiquity. When three comets appeared in succession toward the end of 1618, they generated quite a sensation. The third one, in particular, was first detected on November 27, and around mid-December it became exceptionally impressive, with a long, spectacular tail. Historically, many considered comets to be bad omens, supposedly forewarning of deaths of kings or of bitter wars. As fate would have it, the appearance of the comets did coincide roughly with the beginning of the devastating Thirty Years' War in Central Europe, which resulted in no fewer than 8 million fatalities.

While Galileo may have intended to keep a low profile after the disturbing events against Copernicanism in 1616, it was clear that the advent of the comets would not allow him to remain silent for much longer. At first, Galileo couldn't comment directly on the comets, since he had been bedridden with pain for the entire period during which they were visible and hence unable to observe them with his own eyes. The situation became thornier when a Jesuit mathematician of the Collegio Romano, Orazio Grassi, published in 1619 the contents of his public lecture on the topic, entitled *An Astronomical Discussion on the Three Comets of 1618*.

Grassi, who was a highly educated scientist, an opera stage designer, and an architect, replaced Grienberger as the chair of mathematics in 1617. Like Scheiner before him, Grassi published his treatise anonymously—again, for fear of any potential embarrassment to the Jesuits, in case his ideas turned out to be wrong. Grassi's theory of comets deviated courageously from the Aristotelian view, which placed the comets at about the distance of the Moon. Instead, following Tycho Brahe, Grassi proposed that the comets were considerably farther out, somewhere between the Moon and the Sun. He based this conclusion on "an established law according to which the more slowly they move, the higher they are, and since the motion of our comet was midway between that of the sun and of the moon, it will have to be placed between them." Grassi still adhered to a scheme in which the comets, the Moon, and the Sun orbited the Earth. Brahe's original idea about the distance of comets, by the way, was based on the nondetection of any appreciable parallax (shift with respect to background stars) in the observations of the comet of 1577.

As to the actual nature of comets, many astronomers at the time were still adopting Aristotle's theory, which stated that these represented exhalations of the Earth that became visible above a certain height due to combustion, disappearing from view as soon as that inflammable material was exhausted. Grassi, however, again followed Brahe in suggesting that comets were some sort of "imitation planets." In this, as it turned out, Grassi was more in the right than Galileo, who later defended a view in which comets represented optical effects rather than real objects.

Galileo was alerted to Grassi's publication in the first part of 1619. Even though Galileo's name was never mentioned in the treatise, nor was there anything even remotely offensive to him in it, Galileo was also informed that both the Jesuits of the Collegio Romano and an influential group of Roman intellectuals that included Francesco Ingoli—the person who drew up the Church's modifications to Copernicus—were using the treatise to argue against Copernicanism. Ingoli was employing Brahe's old argument that if the Earth was really moving around the Sun, observations taken six months apart should have revealed a parallax in the position of any celestial object, resulting from the Earth's motion. In the absence of such a detected parallax, Ingoli concluded: "From the motion of the comet, it seems possible not only to refute the Copernican theory, but also to draw forth arguments, whose efficacy is not to be disdained, in favor of the stability of the Earth."

Faced with this open attack on Copernicanism on several fronts at a time when he was still extremely bitter toward the Jesuit mathematicians for deserting him, but also encouraged to intervene by a few of his correspondents in Rome, Galileo decided to respond. Still, understanding the risks involved, he presented his comments not under his own name; instead, he had a former student and recently appointed consul of the Florentine Academy, Mario Guiducci, speak on his behalf. Accordingly, Guiducci gave a series of three lectures about the comets in Florence, and the lectures were published as an essay entitled *Discourse on the Comets of Guiducci*, at the end of June 1619.

Even a cursory examination of this manuscript in the late nineteenth century by Antonio Favaro, the editor of the *National Edition of Galileo's Works*, revealed that it had largely been written (and the rest revised) by Galileo's own hand. While he did not use his most offensive language in the "Discourse" itself, Galileo's annotated copy of Grassi's lectures contains quite a few ill-tempered insults such as *"pezzo d'asinaccio"* ("piece of utter stupidity"), *"bufolaccio"* ("buffoon"), *"elefantissimo"* ("most elephantine"), and *"baldordone"* ("bumbling idiot"). Specifically, Galileo, posing as Guiducci, addresses in the "Discourse" several points. First, he questions whether one could really use parallaxes for comets at all, since it was not obvious at the time that comets indeed represented solid bodies rather than optical phenomena caused by the reflection of light in vapors (similar to rainbows, aurorae, or haloes). Galileo notes:

"There are two kinds of visible objects, the first are true, real, individual and immutable, while the others are mere appearances, reflections of light, images and wandering simulacra. These are so dependent for their existence upon the vision of the observer that not only do they change position when he does, but I believe that they would vanish entirely if his vision were removed."

Another argument in Grassi's publication attracted opprobrium from Galileo. Grassi wrote: "It has been discovered by long experience and proved by optical reasons that all things observed with this instrument [the telescope] seem larger than they appear to the naked eye, yet according to the law that the enlargement appears less and less the farther away they are removed from the eye, it results that fixed stars, the most remote of all from us, receive no perceptible magnification from the telescope. Therefore, since the comet appeared to be enlarged very little, it will have to be said that it is more remote from us than the moon." Grassi appeared to cite here a "law," according to which the magnification of the telescope depended on the distance of the object.

Unfortunately, no such law exists. Galileo, who took all things personally, seems to have thought that this remark was questioning his own understanding of the telescope. Naturally, he wouldn't let it pass without a response. The astronomer who had the best understanding of the optics of the telescope at the time was Kepler. The magnification of a telescope is determined only by the focal lengths—the distance behind the lens over which rays of light that strike it parallel to its central axis are focused—of the objective lens and the eyepiece. The fact that Grassi, who later wrote extensively about optics and who had read Kepler's book, would make such a statement is somewhat puzzling, and it indicates that perhaps he just misspoke.

Even though Galileo's understanding of optics was not of the highest caliber—for example, he confused an increase in the size of the image with the formation of an out-of-focus image—he correctly attacked Grassi's law. Galileo pointed out that if the law were true, one could determine the distance to objects on Earth by merely checking how much the objects were magnified when viewed through the telescope, which was obviously wrong. For example, two telescopes of different power would have shown different magnifications for the same object.

Galileo also took issue with Tycho's original suggestion that comets moved in circular orbits, proposing instead that a motion along a straight line away from the Earth fitted the observations better for the third comet of 1618. We know now that comets indeed move along highly elongated elliptical orbits, locally more resembling motion along a straight line than in a circle.

Galileo never advanced a genuine theory on the nature of comets. In his survey of past ideas, Guiducci/Galileo mentioned favorably the suggestion that comets could represent mere reflections of sunlight by vapors rather than real objects, but he added, "I do not say positively that a comet is formed in this way, but I do say that just as there are doubts about this, so there are doubts about the other schemes employed by other authors." Since Guiducci/Galileo did, however, question the idea that comets represented solid objects, Grassi was justified in concluding that Galileo believed comets to be reflections of sunlight from vapor, with those vapors being exhalations from Earth, rising into the skies. Whereas this hypothetical model was very close to Aristotle's ideas, we should note that Galileo did differ from Aristotle in two important aspects: First, the source of the comet's light was reflected sunlight rather than Aristotle's suggestion of a fiery combustion. Second, Galileo claimed specifically that the comets were far beyond the Moon and therefore well into Aristotle's "celestial" region, which was supposed to be inaccessible to "terrestrial" vapors.

"Discourse on the Comets" was not one of Galileo's best scientific works. Not only was he unable to produce even a provisional, viable theory of comets, but also it contained an inexplicable inconsistency, or internal contradiction. The discrepancy involved Galileo's treatment of the question of the parallax. On one hand, Galileo wanted to refute Grassi's claim that the nondetection of a semiannual parallax in the comets proved that the Earth was not moving around the Sun.

To this end, he argued that one could not really apply parallaxes to comets, since they did not appear to be solid bodies, as evidenced by the fact that one could observe stars through the comets' tails. On the other hand, Galileo did not hesitate to use "the smallness of the parallax observed with utmost care by so many excellent astronomers" to infer the comets' supralunar distances. The fact that these incompatible arguments escaped Galileo's notice is extremely surprising, and Grassi pounced on it in his response to Guiducci's treatise.

There was another serious problem with Galileo's ideas about comets, one which he did realize and remarked upon (via Guiducci):

> I shall not pretend to ignore that if the material in which the comet takes form had only a straight motion perpendicular to the surface of the earth, the comet should have seemed to be directed precisely toward the zenith, whereas, in fact it did not appear so, but declined toward the north. This compels us either to alter what was stated, even though it corresponds to the appearances in so many cases, or else to retain what has been said, adding some other cause for the apparent deviation.

The last sentence alluded to the fact that due to the prohibition on discussing Copernicanism, Galileo didn't feel free to speak his mind. Galileo/Guiducci added that "we should be content with that little bit that we can conjecture *amidst the shadows*, until we are told the true constitution of the parts of the world, because what Tycho had promised remained imperfect." In other words, Galileo recognized that even his hypothetical scenario did not quite agree with observations, but he thought that Brahe's theory was undermined by its own set of challenges. For instance, Brahe suggested that comets move in the opposite direction to that of the other planets. At the same time, Galileo felt formally forbidden to discuss any potential remedies to the model he had examined that could perhaps be provided by the Copernican scenario. In fact, in two works published in 1604 and 1619, Kepler suggested that comets move along straight lines, with the apparent deviation in their path being caused by the Earth's motion. While Galileo was almost certainly inspired by Kepler's ideas, he remained utterly silent about them.

Today we know that comets are small solar system bodies that orbit the Sun along orbits that are either very elongated (highly eccentric) ellipses or hyperbolic. They consist of nuclei that range in size from a few hundred yards to tens of miles across, and are composed primarily of ice, rock, and dust (they are "dirty snowballs"), as well as frozen carbon dioxide, methane, and ammonia. When the comets pass closer to the Sun, solar radiation vaporizes volatile material, which starts streaming out of the nucleus, forming an extended atmosphere, or coma, and two tails, one of dust and one of gas. The dust tail reflects sunlight directly, while the gas tail glows as it is being ionized. The ion tail can be as long as the distance between the Earth and the Sun. In the solar system, there are two reservoirs of comets. One is the Kuiper belt, a disk of comets just beyond Pluto that supplies most of the comets that orbit the Sun with orbital periods of less than a century. The second source, the Oort cloud surrounds the outer solar system, and its outer edge reaches almost a quarter of the way to the nearest star. The Oort cloud may contain as many as a trillion comets, and it supplies the long-period comets. Halley's Comet, arguably the most famous comet, returns to Earth's vicinity about every seventy-five years. It was last seen in 1986.

Galileo was correct in associating comets with the process of releasing gas, and their light with effects triggered by the proximity of the comets to the Sun, but he was wrong in even hypothesizing that the gases were released by the Earth. We should remember, though, that Galileo's goal was not to formulate a definitive theory of comets but mainly to discredit and raise doubts about the model of Tycho Brahe, whose scenario for the solar system he had always regarded as a silly and irritating compromise.

Galileo's ultimate objective was, of course, to refute the Jesuit claim that comets proved Copernicus wrong. In attempting to achieve that, however, he brought upon himself (nobody doubted that Guiducci's publication was Galileo's handiwork) the fury of Orazio Grassi, who complained about Galileo "vilifying the good name of the Collegio Romano," of the Jesuits in general ("The Jesuits are much offended," Giovanni Ciampoli informed him), and even of Scheiner personally, whose work on sunspots was unfavorably and unnecessarily mentioned in the treatise. The arena was thus set for the second round.

GRASSI'S COUNTERATTACK

Guiducci's treatise appeared at the beginning of the summer of 1619, and Grassi took only about six weeks to respond. His stinging, harsh essay, entitled "The Astronomical and Philosophical Balance," hit the press in the fall of the same year. However, this continued to be a battle between two masked men. Just as Grassi's original "Astronomical Discussion" was published with the declaration that it had been authored by "one of the Fathers of the Collegio Romano," and Galileo's response appeared as if it was the work of Guiducci, the *Balance* was published under the somewhat defective (and rather transparent) anagrammatic pseudonym Lothario Sarsio Sigensano, instead of Oratio Grassi Salonensi, with "Sarsi" pretending to be a student of Grassi's. The title contained the word *Balance* because it purported to carefully weigh Galileo's opinions.

Initially, Galileo refused to believe that the *Balance* was written by Grassi, especially because of its sharp sarcasm pointed directly at Galileo. Blind to his own deficiencies when it came to manners and the treatment of others, Galileo felt that this attack was unwarranted, since Guiducci never mentioned Grassi by name. His doubts were quickly dispelled, however, by a letter he received at the beginning of December from his Lincean friend Ciampoli: "I see that you cannot bring yourself to believe that Father Grassi is the author of 'The Astronomical Balance,'" he wrote, "but I repeat to assure you once again that His Reverence and the Jesuit Fathers want you to know that it is their work, and they are so far from the judgment that you

Figure 1. The *Tribuna di Galileo* in the Science Museum "La Specola," Florence.
The statue of Galileo is by Aristodemo Costoli. The frescoes were painted by
Luigi Sabatelli (1772–1850). From left to right they show Galileo observing the
lamp in the Cathedral of Pisa, Galileo presenting his telescope to the Venetian
Senate, and the old and blind Galileo conversing with his disciples.

Figure 2. The earliest known portrait of Galileo, from the last decade of the sixteenth century. Painted by an unknown Tuscan artist. Painting is in the private collection of Florentine art collector Alessandro Bruschi.

Figure 3. Two of Galileo's original telescopes. He designed and constructed these telescopes in his own workshop.

Figure 4. One of Galileo's telescope objective lenses (at the center),
which he produced in late 1609 to early 1610 and used for many
observations. The frame was made by Vittorio Crosten in 1677.
Located at the Galileo Museum, Florence.

Figure 5. Detail of the fresco in the Pauline Chapel of the Basilica of Santa
Maria Maggiore in Rome. Painted by Cigoli (Lodovico Cardi; 1559–1613).
The Moon on which the Virgin stands is pockmarked with craters,
as it had been revealed in Galileo's observations.

Figure 6. "Earthrise," a photo of the Earth and a part of the lunar surface, taken on December 24, 1968, aboard Apollo 8 by astronaut Bill Anders from lunar orbit. Galileo was the first to show that the lunar surface was rugged, like the face of the Earth, and also the first to understand that light reflected from the Earth brightens the lunar night.

Figure 7. Draft of a letter by Galileo to Leonardo Donato, Doge of Venice, with notes on the satellites of Jupiter. This is the earliest surviving record of Galileo's observations of Jupiter.

Figure 8. Galileo at an old age with his disciple and biographer Vincenzo Viviani. Viviani assisted Galileo from 1639 till Galileo's death in 1642 while he was under house arrest at his villa in Arcetri, near Florence. Painting by Tito Lessi (1858–1917).

Figure 9. A bust of Galileo by Giovanni Battista Foggini (1652–1725) on the facade of Viviani's house in Florence. Two large inscriptions glorifying Galileo's life flank the front entrance and give the house its current name: Palazzo dei Cartelloni, *Cartelloni* meaning "Billboards." Viviani turned the facade of his house into a memorial for Galileo.

Figure 10. Index finger, thumb of right hand, and a tooth of Galileo, which were detached from his body when his remains were moved in 1737. These remains are housed in the Galileo Museum.

Figure 11. The author in front of Galileo's tomb at the Basilica di Santa Croce, Florence. The tomb is located opposite that of Michelangelo. It was designed by Giulio Foggini.

Figure 12. Bust of Galileo by Carlo Marcellini (1644–1713),
Galileo Museum, Florence.

make about it that they glory in it as a triumph." Ciampoli added that Grassi himself had usually spoken about Galileo in a much more reserved manner than the other Jesuits did, and that therefore he [Ciampoli] was rather surprised to see Grassi use "so many biting jokes." As we shall see later, judging from Grassi's subsequent behavior, it is hard to avoid the impression that Grassi's "more reserved manner" may have been nothing but an act, to conceal more sinister intentions.

Grassi's *Balance* contained a few valid criticisms. For instance, he pointed out the self-contradiction concerning the absence of a detected parallax, which Galileo used to infer the comets' distances while separately arguing through his avatar Guiducci that parallaxes could not be applied to comets. Grassi also remarked that a few of the ideas advanced by Galileo in Guiducci's "Discourse" were, in fact, not original but rather closely resembled those expressed by the sixteenth-century polymath Gerolamo Cardano and the philosopher Bernardino Telesio. In general, Grassi demonstrated an excellent command of optics and an up-to-date familiarity with all the relevant scientific publications. This was hardly surprising, since even from the little known about him, Grassi was exceptionally knowledgeable. He not only experimented with and wrote about optics and vision, about the physics of light, and about atmospheric pressure, but he was also a great architect who designed the Church of St. Ignatius at the Collegio Romano, as well as a church in Terni and a Jesuit college in Genoa. He even staged an opera, in addition to his accomplishments as a mathematician.

On the other hand, Grassi's essay came with its own set of problems. First, it included a surprisingly naïve reliance on ancient, imaginary tales. Second, it contained an internal self-inconsistency, and third, it directed some underhanded jabs at Galileo. For example, in trying to prove Aristotle's claim that friction with the air could heat bodies to incandescence (which is true in the case of meteorites and artificial satellites reentering Earth's atmosphere), Grassi seemed to trust bizarre stories from antiquity, such as one describing how the Babylonians cooked their eggs by swirling them on slings.

Astonishingly, the inconsistency in *Balance* was associated with the declared purpose of the essay itself. Grassi wrote: "I wish to say that here my whole desire is nothing less than to champion the conclusions of Aristotle." This was a strange assertion, given that his own theory located the comets far beyond the Moon, contrary to Aristotle's notion of immutable heavens. It may be that the insertion of this statement, endorsing Aristotle unequivocally, reflected advice from higher-up Jesuit circles rather than Grassi's own intentions. Finally, there were the "biting jokes," the slyest of which changed Guiducci's phrase "some other *cause* for the apparent deviation" [of the comet's path from the zenith toward the north] to read "some other *motion*." Grassi then wrote this viciously cunning paragraph:

> What is this sudden fear in an open and not timid spirit which prevents him from uttering the word that he has in mind? I cannot guess it. Is this other motion which could explain everything and which he does not dare to discuss—is it of the comet or of something else? It cannot be the motion of the circles, since for Galileo there are no Ptolemaic circles. I fancy I hear a small voice whispering discreetly in my ear: the motion of the Earth. Get thee behind me thou evil word, offensive to truth and to pious ears! It was surely prudence to speak it with bated breath. For, if it were really thus, there would be nothing left of an opinion which can rest on no other ground except this false one.

Then, delivering a parting blow, Grassi added a line strikingly similar to Marc Anthony's recurring, sarcastic pronouncement "And Brutus is an honorable man" in William Shakespeare's *Julius Caesar*: "But then certainly Galileo had no such idea, for I have never known him otherwise than pious and religious."

How can we reconcile these insidious remarks with Ciampoli's statement that Grassi had always spoken about Galileo with respect? A few Galileo scholars suggested that maybe these passages were the work of Scheiner, whose well-known animosity toward Galileo was constantly increasing. At any rate, Galileo had to consider how to react without worsening his relations with the mathematicians of the Collegio Romano. For his part, Mario Guiducci, whose name was, after all, listed as the author of the *Discourse*, responded to Grassi's *Balance* by sending a letter to his former professor of rhetoric at the Collegio Romano, Tarquinio Galluzzi. He did not attempt to address the physical arguments, declaring only that while he had different views on comets from those of the "Reverend Mathematician," he had no intention of offending Father Grassi or any other of the Jesuit mathematicians.

As for Galileo himself, after consulting with his Roman friends Cesi, Ciampoli, Cesi's cousin (and Lincean Academy member) Virginio Cesarini, and another founding member of the academy, Francesco Stelluti, it was decided that Galileo should send his response to Cesarini. In addition to not wanting to further muddy the water, Galileo's friends judged that it would be improper for him to answer directly to Grassi, since the latter had chosen to hide behind the fictional disciple Sarsi.

Virginio Cesarini was an excellent choice to receive Galileo's manuscript, since he was known to be a true believer (he was later chamberlain to two Popes) and also an informed intellectual and a poet, who served as a conduit between scientists working in different cities. His open-mindedness and intellect were perfectly demonstrated in a letter he sent to Galileo in 1618, in which he urged the master to create and spread a new logic "based on natural experiments and on mathematical demonstrations," since he believed that such a "more certain logic . . . will at once open the intellect to consciousness of the truth and shut the mouths of some vain and pertinacious philosophers whose science was opinion and, what is worse, other people's and not their own."

These very sentiments were echoed in a pronouncement by philosopher and mathematician Bertrand Russell more than three hundred years later: "Philosophy is to be studied, not for the sake of any definite answers to its questions . . . but rather for the sake of the questions themselves; because these questions enlarge our conceptions of what is possible, enrich our intellectual imagination, and diminish the dogmatic assurance which closes the mind against speculation."

As it turned out, recurring illnesses, preoccupation with his literary interests, and a series of historically consequential events delayed Galileo's response until October 1622, when he finally sent a manuscript to Cesarini. On the literary side, Galileo returned to his lifelong, almost obsessive fascination with comparing poets Ariosto and Tasso. The significant historical incidents included the deaths of Pope Paul V, Cardinal Roberto Bellarmino, and, even more impactful for Galileo, his greatest supporter, the Grand Duke Cosimo II, all in 1621. Since Cosimo's son Ferdinando was only ten years old at the time of his father's death, the Grand Duchess Christina and her daughter-in-law, Maria Maddalena of Austria, both singularly religious women, were appointed as regents of the duchy.

Galileo's manuscript, when eventually completed, was entitled *The Assayer*, referring to an extremely precise scale that goldsmiths were using, and providing a contemptuous contrast to Grassi's *Balance*, which implied a cruder weighing device. As soon as Cesarini received the manuscript, he sent copies of it for comments to Ciampoli, Cesi, and a few other friends. He also informed Galileo that the Jesuit mathematicians, who had heard about the manuscript's arrival, were "eager and anxious, and they even dared to ask me for it; but I have refused them it because they would have been able more effectively to obstruct its publication."

In spite of these assurances, Cesarini did not resist the temptation to read parts of *The Assayer* to a few of his acquaintances. One way or another, the Jesuits also heard about those passages—in seventeenth-century Rome, they were the equivalent of Big Brother—and, according to Cesarini, "they have fathomed everything." There was still the thorny problem of obtaining permission to print the pamphlet. At the time, it was standard practice that a "print approval," or imprimatur, would have to be obtained from the ecclesiastical authorities for any manuscript to be published. Cesarini managed to have the book examined by the Dominican Niccolò Riccardi, who was a Genoese follower of Galileo. Riccardi did not disappoint. He expressed effusive admiration for the book: "thanks to the subtle and solid speculation of the author in whose days I consider myself happy to have been born, when, no longer with the steelyard [a weighing apparatus] and roughly, but with such delicate assayers the gold of truth is weighed." This was not the common imprimatur that the Inquisition's bureau-

Figure 8.1. Title page of *The Assayer*.

cracy expected. It sounded more like present-day laudatory blurbs that appear on the covers of new books. Cesarini was happy to take it. He rapidly inserted the revisions suggested by about a half dozen members of the Accademia dei Lincei and rushed the work to print.

Still, unexpected external circumstances further delayed publication. On July 8, 1623, Pope Gregory XV died after only two years of papacy. Then, following exhausting negotiations among the cardinals, Cardinal Maffeo Barberini was elected on August 6 as Pope Urban VIII. After years of the nonerudite (although surprisingly undogmatic in doctrinal matters) Pope Paul V, the election of a relatively young, brilliant, presumably open-minded, and refined intellectual Pope was greeted with hope by Galileo, his friends, and, indeed, all progressive Catholics. As a cardinal, Barberini had shown great admiration for Galileo, even to the point of sending him an ode, "Adulatio Perniciosa," in which he expressed his esteem for Galileo's astronomical discoveries. Barberini had also apparently played a role in preventing Copernicanism from being pronounced altogether heretical in 1616. Perhaps most important, shortly before Barberini was elected as Pope, Galileo congratulated him on the occasion of his nephew, Francesco Barberini, having completed his studies. In his reply, Barberini wrote: "assuring you that you find in me a very ready disposition to serve you out of respect for what you so merit and for the gratitude I owe you."

Given these expressed sentiments and the fact that Pope Urban VIII appointed Galileo's friends Cesarini, Ciampoli, and Stelluti to master of the chambers, secretary of the briefs, and privy chamberlain, respectively, it should come as no surprise that the Accademia dei Lincei dedicated *The Assayer* to the Pope. In October 1623 the book was finally ready. Unfortunately, it still contained many typographical errors, but it included a wonderful dedication written by Cesarini himself and signed by all the members of the Lincean Academy. The dedication read, in part: "This we dedicate and present to Your Holiness as to one who has filled his soul with true ornaments and splendors and has turned his heroic mind to the highest undertakings. . . . Meanwhile, humbly inclining ourselves at your feet, we supplicate you to continue favoring our studies with gracious rays and vigorous warmth of your most benign protection."

Cesi, the founder of the academy, presented a splendidly bound copy to the Pope on October 27. (Figure 8.1 shows the title page.) Copies of the book were also given to the cardinals. This marked the official approval of *The Assayer*, a book whose literary verve and intellectual passion were hailed by one of Galileo's twentieth-century biographers as "a stupendous masterpiece of polemic literature." Grassi's opinion was, of course, very different.

CHAPTER 9

The Assayer

Even though *The Assayer* was formally written as a response to Orazio Grassi's (Sarsi's) *Balance*, the topic of comets became rather peripheral in Galileo's polemical masterpiece, providing in some sense only a pretext for an exposition of Galileo's thoughts about various aspects of science, and a platform for him to attack Tycho Brahe's system.

Right from the start, Galileo presents two of his principal viewpoints: first, his disdain for blind reliance on authority and, second, his philosophy on the nature of the cosmos. The following passage has become one of Galileo's most memorable manifestos:

> In Sarsi I seem to discern the firm belief that in philosophizing one must support oneself upon the opinion of some celebrated author, as if our minds ought to remain completely sterile and barren unless wedded to the reasoning of some other person. Possibly he thinks that philosophy is a book of fiction by some writer, like the *Iliad* or *Orlando Furioso*, productions in which the least important thing is whether what is written is true. Well, Sarsi, that is not how matters stand. Philosophy is written in this grand book, the universe, which stands continually open to our gaze. But the book cannot be understood unless one first learns

to comprehend the language and read the letters in which it is composed. It is written in the language of mathematics, and its characters are triangles, circles, and other geometric figures without which it is humanly impossible to understand a single word of it; without these, one wanders about in a dark labyrinth.

Galileo's proclamation about the mathematical nature of reality is particularly striking. We should remember that he made this statement at a time when very few mathematical "laws of nature" had been formulated (mostly by him!). Yet he somehow anticipated what Nobel laureate Eugene Wigner would call in 1960 "the unreasonable effectiveness of mathematics"—the fact that the laws of physics, which the entire universe seems to obey, are all stated in the form of mathematical equations. Even earlier, in 1940, Einstein turned this fact into the definition of physics: "What we call physics comprises that group of natural sciences which base their concepts on measurement, and whose concepts and propositions lend themselves to mathematical formulations." But what is it that gives mathematics such powers?

With very little evidence to base this opinion upon, Galileo thought in 1623 that he knew the answer: the universe "is written in the language of mathematics." It was this dedication to mathematics that raised Galileo above Grassi and the other scientists of his day, even when his specific arguments fell short of convincing—and even though he assigned to geometry a more important role than it seemed to deserve at the time. His opponents, he wrote, "failed to notice that to go against geometry is to deny truth in broad daylight."

Impressively, together with his conviction that nature was geometrically structured, Galileo also understood that all scientific theories are only tentative and provisional. That is, science has to be constantly reappraised as fresh observational evidence becomes available. By admitting that everything he said had been "set forth tentatively as a conjecture . . . open to doubt and, at best, only probable,"

Galileo introduced the revolutionary departure from the medieval, ludicrous notion that everything worth knowing was already known. Instead, Galileo expressed only one near certainty: that to decipher any of nature's secrets would require the language of mathematics.

The Assayer offered its author the opportunity to exhibit some of his most witty sarcasm. For example, he and Grassi differed in their understanding of the origin of heat. Whereas Grassi, following the Aristotelians, thought that heat was caused entirely by motion, Galileo attributed heat also to the detachment of matter particles via friction forces, or to compression. In modern terms, heat is a form of energy transferred, for example, due to a temperature difference between two systems, with the temperature being determined by the average speed of the random motion of atoms or molecules. The problem with Grassi's concept was that because of his trust in ancient authors, he committed the naïve error of believing legendary tales, such as that mentioned earlier of the Babylonians cooking eggs through whirling them on slings. Galileo pounced on this fallacy like a cat on a slow mouse:

> If Sarsi wishes me to believe, on the word of Suidas [Greek historian], that the Babylonians cooked eggs by whirling them rapidly in slings, I shall believe it; but I shall say that the cause of this effect is very far from the one he attributes to it. To discover the true cause, I reason as follows: "If we do not achieve an effect which others formerly achieved, it must be that we lack something in our operation which was the cause of this effect succeeding, and if we lack one thing only, then this alone can be the true cause. Now we do not lack eggs, or slings, or sturdy fellows to whirl them, and still they do not cook, but rather cool down faster if hot. And since we lack nothing except being Babylonians, then being Babylonian is the cause of the egg hardening."

Another interesting and consequential discussion in *The Assayer* concerned the very nature of matter and the role of the senses. Following a distinction that dated all the way back to the Greek philosopher Democritus in the fifth century BCE, Galileo identified two types of properties: those that were intrinsic to physical bodies, such as shapes, numbers, and motions, and those that were, in his view, associated with the existence of conscious, sentient observers, such as tastes and odors. He wrote:

"To excite in us tastes, odors, and sounds, I believe that nothing is required in external bodies except shapes, numbers, and slow or rapid movements. I think that if ears, tongues, and noses were removed, shapes and numbers and motions would remain, but not odors or tastes or sounds. The latter, I believe, are nothing more than names when separated from living beings."

Galileo's reintroduction of these concepts from antiquity into the conversation of early seventeenth-century philosophy may have later inspired and affected similar ideas of Descartes and especially of the empiricist philosopher John Locke. In his influential 1689 treatise *Essay Concerning Human Understanding*, Locke specifically distinguished between what he regarded as properties that are independent of any observer (dubbed "primary qualities"), such as number, motion, solidity, and shape, and "secondary qualities": those that produce sensations in observers, such as color, taste, odor, and sound. As we shall see later, even this seemingly innocuous discussion of those qualities that Galileo regarded as being subjective, and representing mere names in the external object, was about to contribute, to some extent at least, to his subsequent troubles with the Church.

When all was said and done, Galileo's main goal in *The Assayer* was, from the outset, to destroy Tycho's scenario, which he regarded as the only remaining obstacle to convincing everybody of the truth of the heliocentric model. Indeed, in his *Balance*, Grassi (speaking as Sarsi) defended his use of Tycho's parallax argument:

Let it be granted that my master followed Tycho. Is this such a crime? Whom instead should he follow? Ptolemy? Whose followers' throats are threatened by the out-thrust sword of Mars now made closer. Copernicus? But he who is pious will rather call everyone away from him and will spurn and reject his recently condemned hypothesis. Therefore, Tycho remains as the only one whom we may approve of as our leader among the unknown course of the stars.

The point about Copernicus was a pitiable argument to bring into a scientific debate—one that demonstrated precisely how convictions entirely irrelevant to the subject matter could color and distort opinions. Grassi relied here on the nonscientific 1616 decree against Copernicanism to argue that one couldn't even *consider* the Copernican model of the solar system. Unfortunately, similar attitudes continue to sometimes dominate thought to this very day. A policy today that encourages the teaching of creationism as thinly veiled "intelligent design," for instance, in order to steer students' minds away from Darwin's theory of evolution, amounts to precisely the same practice.

This is not to say that Galileo was always right. In fact, as I noted earlier, his specific arguments about comets contained two glaring inconsistencies: one, in claiming that parallaxes couldn't be used for comets—only to then turn around and use them to determine the distance to comets; and the other, in suggesting that comets move along a straight line—only to admit later that, in fact, they don't. These were scientific errors that Grassi correctly pointed out and criticized. Science is not infallible. On the contrary, Galileo himself recognized that every scientific theory is subject to confirmation. What only science can promise, however, is a continuous, midcourse self-correction, as additional experimental and observational evidence accumulates, and new theoretical ideas (all based on mathematics, Galileo believed) emerge. Even with all his understandable caution to avoid being accused of Copernicanism, Galileo couldn't abandon his trust in the burgeoning scientific method, insisting that philosophers should also use "natural reason, when one can," to demonstrate the "falseness of those propositions which are declared to be against Holy Scripture."

As could be expected, *The Assayer* was received rather dissimilarly by Galileo's Roman friends than by Grassi. The latter reportedly rushed to the Sun bookshop, where the first copy had been on display, emerging with a "changed color," the book under his arm. Pope Urban VIII, on the other hand, apparently enjoyed *The Assayer*'s aggressive satire and pointed sarcasm, since he had it read at his table for entertainment.

Grassi, keen on publishing a reply, wrote a new book relatively rapidly (still using the pseudonym Sarsi), entitled *Comparison of the Weights of the* Balance *and* The Assayer." Aware of the Pope's support for Galileo's book, however, he published his book in Paris, which caused a considerable delay in its becoming available. Galileo read *Comparison* but decided that it would be a waste of time to respond yet again, even though the book did contain one worrisome innuendo. The insinuation concerned Galileo's remarks on the subjective nature of qualities such as taste, smell, and color. Grassi claimed that this description went against the Catholic doctrine of the miracle of the Eucharist, which required the preservation of the taste and smell of the bread and the wine, even though their substance was transformed in a way beyond human comprehension into Christ's body, blood, and soul.

Italian scholar Pietro Redondi discovered and published in 1983 a previously unknown document in the archives of the Holy Office. This document, which Redondi attributed to Grassi, denounced Galileo as a heretic. In this new twist, the accusation was based on the fact that Galileo "openly declares himself a follower of the school of [ancient Greece philosophers] Democritus and Epicurus"—meaning that he believes in atoms—a belief that was considered incompatible with the trans-substantiation underlying the dogma of the Eucharist. From this letter, Redondi developed an imaginative conspiracy theory contending that the real heresy for which Galileo was eventually condemned was not Copernicanism but, rather, atomism. While most fellow historians do not accept Redondi's speculation, there is no doubt that the addition of one more item to Galileo's growing list of troubles certainly did not help.

The bottom line was that in spite of the Pope's apparent support and some reassurances from Father Niccolò Riccardi that Galileo's opinions "were not otherwise against the Faith," Galileo felt that he still had serious reasons to be concerned. With respect to atomism, today we believe that all ordinary matter is composed of some elementary particles that are not composed of other particles. In the extremely successful Standard Model of particle physics, those elementary particles include quarks (of which protons and neutrons are made), leptons (electrons, muons, and neutrinos), gauge bosons (force carriers), and the Higgs boson (which is an excitation of a certain field). Everyday matter is indeed composed of atoms, once presumed to be elementary but known today to contain those subatomic elementary particles.

All the worries, anxiety, and fears associated with Grassi's attacks notwithstanding, the relative success of *The Assayer* must have given Galileo some satisfaction. His conviction in the correctness of the Copernican model was too strong for him to give it up at that stage.

CHAPTER 10

The *Dialogo*

The Assayer did not present Galileo at his best as a scientist. Rather, it showed him more as a magician with words and tricky logic, and demonstrated his brilliance and articulation as a debater. However, the election of Maffeo Barberini as Pope Urban VIII, resurrected Galileo's hopes that he could perhaps change the Church's position vis-à-vis Copernicanism. With this goal in mind, Galileo wanted to meet with the Pope as early as possible, but his poor health prevented him from traveling to Rome until the spring of 1624. Pope Urban graciously granted Galileo no fewer than a half dozen audiences and showed him great respect and generosity, but the practical results fell short of Galileo's expectations. He emerged from those meetings realizing that in spite of the new Pontiff's open-mindedness, Urban VIII was convinced that humans would never be able to comprehend the mysteries of the cosmos. To the Pope, irrespective of which theory of planetary motions scientists were to adopt, "we cannot limit the divine power and wisdom to this way." Galileo's views were, of course, very different. Neverthe-less, he came out with the impression that he was allowed to present the Copernican model as a *hypothesis* and to show that on scientific grounds, at least, this conjecture explained the observations better than the Aristotelian-Ptolemaic systems. As he was about to find out, even this impression was inaccurate.

Upon his return to Florence, Galileo decided to proceed step-by-step, first by answering Francesco Ingoli's eight-year-old publication against Copernicanism. Ingoli was the person who "corrected" Copernicus's book for the Congregation of the Index. Galileo's tactic was to present himself as choosing not to be a Copernican "for higher motives" [that is, for being a pious Catholic] rather than for scientific reasons, even though science was the area in which he exposed Ingoli's arguments as very weak if not downright false. This was a risky move, and upon Federico Cesi's advice, the letter to Ingoli was never delivered because "Copernicus's opinion is explicitly defended, and though it is clearly stated that this opinion is found false by means of a superior light," Cesi estimated that there would be those who "will not believe that and will be up in arms again." Galileo's close friend was undoubtedly right, since other events unfavorable to Galileo were piling up around that time. Perhaps the most significant were the premature death of his great supporter Virginio Cesarini, and the fact that Cardinal Alessandro Orsini, a previously enthusiastic admirer of Galileo's, had joined the Jesuit order and had become greatly influenced by Galileo's sworn enemy Christoph Scheiner. In addition, Mario Guiducci informed Galileo that the Holy Office had received a proposal from an unidentified person to add *The Assayer* to the list of prohibited books because of its pro-Copernican contents.

All of this did not stop Galileo from starting around 1626 to compose his next major book, the *Dialogo*, which originally was supposed to describe in detail his theory of the tides—a phenomenon he still regarded as the most convincing proof for the Earth's motion. The work, however, progressed very slowly over the next three years and with prolonged interruptions caused sometimes by Galileo's declining health and sometimes by his needing more data on tides. Somewhat surprisingly, perhaps, Galileo decided not to react in any substantial way (except for a disgruntled letter to Prince Paolo Orsini) to Scheiner's massive work on sunspots, *Rosa Ursina*, published in 1630.

By the time Galileo was putting the final touches on the *Dialogo*, Lady Luck appeared to show him some kindness: Father Niccolò Riccardi, who had glowingly commended *The Assayer* some years earlier, was appointed master of the Sacred Apostolic Palace in June 1629. With that title, Riccardi became the person who gives the final authorization to print. Consequently, Galileo's friends were cautiously optimistic about being able to print the book in Rome. Castelli, who now had an appointment as a professor of mathematics at the University of Rome, wrote to Galileo that Ciampoli "holds as solid" that were Galileo to come to Rome with the work in hand, he would "overcome whatever difficulty" he might meet.

Galileo arrived in Rome on May 3, 1630, and was received as an honored guest by the Tuscan ambassador, Francesco Niccolini, who had been appointed to this post in 1621. About two weeks later, he was granted an audience with Urban VIII. Undoubtedly the Pope reiterated his previous opinions about the need to treat Copernicanism merely as a hypothesis and his belief that the universe would always remain beyond human comprehension. Nevertheless, based on the Pope's general demeanor and warmth, Galileo apparently convinced himself that the Pontiff would not object to the publication of what was to become the *Dialogo*.

Galileo did fail, however, to appreciate two crucial facts. The first had to do with the sensitive political situation and psychological state of the Pope himself. Urban VIII, who was a genuine lover and patron of the arts, had spent money extravagantly during his papacy—a splurge culminating in the sumptuous Palazzo Barberini, a seventeenth-century palace in Rome. At the same time, he funded the construction of a variety of fortresses and other military endeavors, financially debilitating a pontificate perceived already as nepotistic and consumed with desires for earthly delights. In addition, the Thirty Years' War had been raging for more than a decade, with no end in sight, and even Rome's relations with France, a country that Urban VIII generally supported, had been somewhat strained by the uncompromising positions adopted by the influential French cardinal

Armand Jean du Plessis Richelieu. All of these predicaments had turned Pope Urban VIII into a moody, capricious, and suspicious man who demanded absolute obedience on all fronts from all of those surrounding him.

The second reality that Galileo failed to recognize fully was the level of hatred a few of his enemies harbored toward him and for new scientific ideas in general, and the cruel steps that they were prepared to take to bring about his downfall. This animosity manifested itself in an appalling incident during Galileo's stay in Rome. The event went like this: the abbot of Saint Praxedes in Rome apparently published a horoscope predicting the imminent death of the Pope and of his nephew. Some of Galileo's adversaries tried to pin the blame on Galileo, announcing:

> Here we have Galileo, who is a famous mathematician and astrologer, and he is trying to print a book in which he impugns many opinions held by the Jesuits. He has let it be known . . . that at the end of June we will have peace in Italy, and that a little afterwards Sir Taddeo and the Pope will die. The last point is supported by Caracioli Neapolitan, by Father Campanella, and by many written discourses, which treat the election of a new Pontiff, as if the See were vacant.

Galileo, who knew how superstitious Pope Urban VIII was, had to react immediately and send word to the Pope denying any involvement in the affair. Fortunately, this particular vicious plot did not succeed, and the Pope assured Galileo that he was clear of any suspicion.

Father Riccardi, who was in charge of approving the *Dialogo*, was perfectly aware of the delicate situation in Rome at the time. After his first reading of the manuscript, he realized at once that despite what Galileo might have thought, and even though the final result of the discussion was left inconclusive, this was, at least in large parts, an unmistakably pro-Copernican text, which could spell serious trouble if published unedited. He therefore suggested that, in addition to some necessary revisions, an introductory section or preface, and a concluding chapter be added, which would emphasize the hypothetical nature of the Copernican model. Consequently, it was decided that both Riccardi himself and the Dominican Raffaele Visconti would review the book thoroughly before discussing it with the Pope. That conversation with the Pontiff eventually took place in mid-June 1630, and based on what had been presented to him (which was partial at best), the Pope expressed his general satisfaction. He did insist, though, on the title to not concentrate on the tides—since that would have implied that the main topic was proof of the Earth's motion—but on the "Chief World Systems." With these assurances and a friendly parting from the Pope and his nephew Cardinal Francesco Barberini, Galileo finally departed for Florence on June 26, 1630.

Unfortunately, this was not the end of the trials and tribulations Galileo had to endure for the publication of the *Dialogo*. Most significant of these was the sudden death on August 1, 1630, of Federico Cesi, the founder and sole source of funding for the Accademia dei Lincei. As a result, the printing had to be done in Florence, outside of Riccardi's jurisdiction, instead of Rome. After some negotiations, it was agreed that Father Jacinto Stefani, a consultor to the Inquisition in Florence, would be in charge, but only after Riccardi approved the introduction and conclusion. The entire operation was painfully slow. Galileo, who by now had lost his forbearance, consented to a meeting with all the Florentine parties of authority involved, and he stated impatiently:

> "I agree to give the label of dreams, chimeras, misunderstandings, paralogisms, and conceits to all those reasons and arguments which the authorities regard as favoring opinions they hold to be untrue; they would also understand how true is my claim that on this topic I have never had any opinion or intention but that held by the holiest and most venerable Fathers and Doctors of the Holy Church."

To make a very long story short, the printing of the *Dialogo* wasn't completed until February 21, 1632. The book listed the permissions (imprimatur) of both Riccardi and of the inquisitor in Florence, Clemente Egidi, even though Riccardi himself had not seen the final version but had sent the instructions about the introduction and ending to Egidi. Respecting the Pope's request, the title read (not including various attributions) *Dialogue Concerning the Two Chief Systems of the World, Ptolemaic and Copernican, Propounding Inconclusively in the Philosophical Reasons as Much for the One Side as*

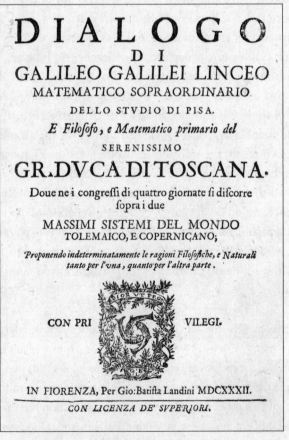

Figure 10.1. Title page of the *Dialogo*.

for the Other (Figure 10.1 shows the title page). There was a certain sleight of hand in the title. Even if one were to ignore the fact that the Aristotelian and the Ptolemaic systems were not identical, there was at least one other world system that in terms of agreement with observations was superior to the Ptolemaic: Tycho Brahe's hybrid system in which the planets revolved around the Sun, but the Sun itself revolved around the Earth. Galileo always regarded that system as unnecessarily complex and contrived, and he also thought that he'd found proof for the Earth's motion through the phenomenon

Figure 10.2. Frontispiece of the *Dialogo*.

of the tides, so in striving to hand Copernicanism a clear victory (although formally the book was inconclusive), he probably didn't want to confuse the issue with superfluous qualifications.

In the all-important preface—added at Father Riccardi's request in order to assist in obtaining permission to print—Galileo did his best to give the impression that he agreed with the 1616 anti-Copernican decree. Readers today may perceive that he barely succeeded in hiding his sarcasm and contempt for the decree and for the anti-intellectual constraints imposed on him personally:

> There were those who impudently asserted that this decree had its origin not in judicious inquiry, but in passion none too well informed. Complaints were to be heard that advisors who were totally unskilled at astronomical observations ought not to clip the wings of reflective intellects by means of rash prohibitions.
>
> Upon hearing such carping insolence, my zeal could not be contained. Being thoroughly informed about that prudent determination, I decided to appear openly in the theater of the world as a witness of the sober truth. I was at that time in Rome; I was not only received by the most eminent prelates of that Court, but had their applause; indeed, this decree was not published without some previous notice of it having been given to me [by Cardinal Bellarmino].

To further please the Pope, Galileo went against his personal scientific convictions and declared that with the *Dialogo*, he had "taken the Copernican side in the discourse, proceeding as with a pure mathematical hypothesis." That is, he pretended to accept the "saving the appearances" approach to science. Finally, he also added a direct reference to the Pope's view that even if the Copernican system explains the motions of the planets, it might not represent reality because God is all powerful and could have created the same appearance by some entirely different means, beyond human understanding. Along these lines, Galileo wrote:

"It is not from failing to take count of what others have thought that we have yielded to asserting that the earth is motionless . . . but (if for nothing else) for those reasons that are supplied by piety, religion, the knowledge of Divine Omnipotence, and a consciousness of the limitations of the human mind." Naïvely, Galileo thought that these disclaimers would be sufficient.

SALVIATI, SIMPLICIO, SAGREDO

The *Dialogo* is one of the most engaging science texts ever written. There are conflict and drama, yes, but also philosophy, humor, cynicism, and poetic usage of language, so that the sum is much more than its parts.

Fashioned after Plato's dialogues, the *Dialogo* was presented as an imaginary discussion among three interlocutors, which takes place in a Venetian palace over a period of four days. Salviati, named for Galileo's deceased Florentine friend Filippo Salviati, is a stand-in for Galileo's Copernican opinions. Sagredo, named for Galileo's great (also deceased) Venetian friend Gianfrancesco Sagredo, plays the role of the educated but nonexpert person who wisely judges between the Copernican and Aristotelian views expressed by the other two. Finally, Simplicio is an avid Aristotelian, who obstinately defends the geocentric worldview. He was supposedly named after Simplicius of Cilicia, the sixth-century commentator on Aristotle's works, with the name being a double entendre, hinting at simplemindedness. Simplicio was fashioned partly after the conservative Cesare Cremonini and partly after Galileo's nemesis, Lodovico delle Colombe.

During the first three days, Galileo's alter ego, Salviati, methodically demolishes Simplicio. Using examples ranging from dead cats falling out of windows to the illusion that the Moon follows us as we wander along a path, Galileo rejects any ancient authorities (such as Aristotle), "for our disputes are about the sensible world, and not one of paper."

In the first day, he demonstrates that there is no difference between terrestrial and celestial properties. In the second day, he intimates that all the observed motions in the heavens are explained more easily by assuming that it is the Earth that moves rather than the Sun and the rest of the world.

Salviati devotes the third day to countering all the objections raised against the Earth's revolution around the Sun and to providing evidence that it indeed moves. Perhaps most interesting in this discussion is Galileo's new claim that he can *prove* the reality of the annual motion of the Earth from the observations of the paths of sunspots on the solar surface. Galileo's and even more so Scheiner's detailed observations of sunspots had revealed that the projected path of sunspots is not along a straight line parallel to the ecliptic. Rather, during one quarter of the year, they appear to ascend in a straight line inclined to the ecliptic. In the following quarter, they move along a path curved upward; in the next one, along a descending straight line; and, in the last quarter, they follow a downward curved path (as shown schematically in Figure 10.3). Galileo demonstrated that the root cause for the curve traced by these apparent motions was a rotation of the Sun about its axis, with a tilt of about 7 degrees of the solar spin axis with respect to a line perpendicular to the ecliptic plane. Relying then on Occam's razor, that of two explanations to a given phenomenon the one that requires fewer assumptions is usually correct (in Galileo's words: "whatever can be accomplished through few things, is done in vain through more"), he further claimed clear superiority of the Copernican system (over the Ptolemaic one) in explaining these observations. Since Galileo apparently hit upon this

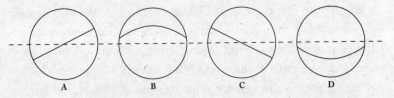

Figure 10.3. Schematic showing the directions of the observed paths of the annual motion of sunspots, in four quarters.

particular proof only a few months before submitting the *Dialogo* to press, his explanations are rather vague and certainly insufficient, which has caused many Galileo scholars to be skeptical about the proof's validity. (Hungarian British author Arthur Koestler even went so far as to lambaste Galileo for both stupidity and dishonesty.)

More recent, thorough analyses of the proof, however, showed that while Galileo did not include all the relevant motions, the paths of sunspots could indeed be used as convincing evidence in favor of the Copernican system. More important, perhaps, even though Galileo hadn't realized it, his proof militated as decisively against Tycho's system as it did against the Ptolemaic scenario. It was certainly much stronger than the proof from the ebb and flow of the tides to which Galileo devoted the fourth day of the *Dialogo*. Interestingly, Galileo was fully aware of the explanations of the tides that relied on the Moon's influence, but in the absence of a theory of gravitation, he regarded ideas such as those of Kepler—who talked specifically about "attractive forces" between the Moon and the Earth—as employing "occult properties," even though Kepler's concept happened to be a genuine forerunner to Newton's theory.

In his summary of the four-day marathon of discussions, Sagredo concludes:

> In the conversations of these four days, we have, then, strong evidences in favor of the Copernican system, among which three have been shown to be very convincing—those taken from the stoppings and retrograde motions of the planets, and their approaches toward and recessions from the earth; second, from the revolution of the sun upon itself, and from what is to be observed in the sunspots; and third, from the ebbing and flowing of the ocean tides.

As we have seen, the third claimed piece of evidence (the tides) was, in fact, incorrect, and the second (the paths of sunspots) may have been an even stronger proof than Galileo realized or was able to articulate. With incredible foresight, Galileo added a fourth test: "The fourth, I mean, may come from the fixed stars, since by extremely accurate observations of these there may be discovered those minimal changes that Copernicus took to be imperceptible." Galileo predicted here that the tiny parallax shift against the background stars due to the motion of the Earth around the Sun would eventually become measurable, as indeed it has.

But, you may wonder, how could Galileo afford to finish his book with an advocacy for Copernicus? After all, the injunction imposed on him in 1616 by Seghizzi explicitly ordered him not to do so. Indeed, he could not. The risk of provoking a harsh punishment by the Church was just too high. Instead, he was forced into finishing with qualifications and reservations essentially negating the entire contents of the book! The renunciation was expressed most clearly by Simplicio:

> I know that if asked whether God in His infinite power and wisdom could have conferred upon the Watery element its observed reciprocating motion [the tides] using some other means than moving its containing vessels, both of you would reply that He could have, and He would have known how to do this in many ways which are unthinkable to our minds. From this I forthwith conclude, this being so, it would be excessive boldness for anyone to limit and restrict the Divine power and wisdom to some particular fancy of his own.

These were, almost verbatim, the words of Pope Urban VIII. To complete this involuntary, unscientific admission, Galileo had Salviati agreeing completely with Simplicio and accepting that "we cannot discover the work of His hands," and that "much less fit we may find ourselves to penetrate the profound depths of His infinite wisdom."

Galileo may have believed that by repeating the Pope's own views on the human inability to truly fathom the cosmos, he'd paid his dues to the anti-Copernican philosophy—and Father Riccardi may have consented, at least to some extent. In doing so, however, Galileo underestimated the zeal of his enemies, who were sure to notice that the admission of the incomprehensibility of the universe was assigned to Simplicio, who had been ridiculed throughout the entire *Dialogo*.

More important, the act of removing humans from their central place in the cosmos was too brutal to be remedied by some philosophical pleasantries at the end of a long debate with a very different tone.

A few current historians of science have raised a different issue with the conclusion of the *Dialogo*. They regarded Galileo's corrective "clarifications" as a sign of duplicity and cowardice. I cannot disagree more. The *Dialogo* courageously expressed Galileo's genuine opinion on a topic that he had been warned not to discuss. There is little doubt that the delicate balancing act in the preface and the conclusion was imposed on him by his friends and those wishing to ensure that the book would be approved for publication. Galileo could have avoided all the misfortunes, grief, and suffering that were about to ensue by simply being less pugnacious and by not publishing the *Dialogo*. But he was only human, after all, and his sense of personal pride in his discoveries and an uncontrollable passion for what he regarded as the truth were too strong for him to just give up. To Galileo, the task of convincing everybody of the correctness of Copernicanism must have taken the shape of a historical duty. This was why he wrote the *Dialogo* (as he did with most of his other books) in Italian rather than Latin, so that it could be read by any literate and interested Italian. He did his best to convey the universe's beauty and rational coherence, but he left the final judgment to the reader, as Salviati very clearly expresses toward the end of the third day:

I do not give these arguments the status of either conclusiveness or inconclusiveness, since (as I have said before) my intention has not been to solve anything about this momentous question, but merely to set forth those physical and astronomical reasons which the two sides can give me to set forth. I leave to others the decision, which ultimately should not be ambiguous, since one of the arrangements must be true and the other false.

History has indeed proved that Galileo was right, but being right is sometimes insufficient. It certainly did not save Galileo from the hardships and anguish that were in store for him the following year.

CHAPTER 11

The Gathering Storm

As one could have expected, Galileo's fans greeted the publication of the *Dialogo* with great enthusiasm, none perhaps more so than Benedetto Castelli, who was not only a good mathematician and a firm Copernican but also a lifelong supporter of his former teacher. However, no more than four months had passed since the book hit the press, when some disturbing news started to arrive.

It began in the form of a letter from Father Riccardi to the inquisitor in Florence, Clemente Egidi, asking the latter to immediately stop the distribution of the book until some corrections were sent from Rome. This ominous act was in itself the consequence of a series of events that accumulated to fuel in the Pope's mind suspicion and hostility toward Galileo.

The first incident involved Galileo's friend Giovanni Ciampoli. By choosing to support Spanish-leaning cardinals, Ciampoli had put himself in the opposition to the Pope's generally Francophile strategy, thus losing Urban VIII's sympathy and trust. In addition, Ciampoli wrote a letter in which he criticized the Pope's style, thereby appending a personal element to the Pontiff's disapproval of him. Second, the Pope was starting to get word, especially from Jesuit opponents of Galileo, that the content of the *Dialogo* was different from what

Urban VIII had expected. In particular, he was made aware of the fact that his own main argument about the incomprehensibility of the universe, and the inability of humans to ever prove the reality of any theoretical world system, had been treated disrespectfully in the *Dialogo*. Not only was it presented rather marginally and very briefly, but to add insult to injury, it was put in the mouth of Simplicio, who had been ridiculed throughout the entire book. Finally, due to the Pontiff's generally paranoid state of mind at that time, Urban VIII even misinterpreted the printer's seal on the book's front page—composed of three dolphins, each holding the other's tail in its mouth (see bottom of Figure 10.1, page 166)—to be an allusion to his own nepotism with his nephews. The upshot of all of these unnerving occurrences was that by August 1632, the talk in church circles in Rome already included options ranging from a delay in the book's distribution, to its outright prohibition. In particular, the Jesuit fathers were reported to be actively trying to ban the book and to "execute him [Galileo] most bitterly." In the meantime, Father Riccardi was trying diligently to obtain information on how many copies of the *Dialogo* had been printed and to whom they'd been sent, in order to recall all of them.

What did Galileo do in response to all of these negative developments? Only the very little he could do: he asked both the Florentine ambassador to Rome, Niccolini, and the grand duke Ferdinando II himself to protest the restrictions imposed on a book that had previously received all the required permissions and licensing. Niccolini indeed met several times with the Pope's nephew Cardinal Francesco Barberini, and upon hearing that a commission composed entirely of people unfriendly to Galileo had been appointed to examine the book, he requested that "some neutral persons" would be added to this commission. The ambassador received no assurances.

In early September Filippo Magalotti, a relative of the Barberini family and a friend of both Galileo and Mario Guiducci, finally heard from Father Riccardi the main complaints against the *Dialogo*. In addition to the diminished role given to the Pope's views in the book, the claim was that the preface was insufficient in providing the necessary balance to the Copernican opinions expressed in the main body of the *Dialogo*—especially since the preface appeared to be, and in fact was, an afterthought. At that stage, Magalotti still expressed cautious optimism that "with some small thing that is removed or added on for greater caution . . . the book will remain free." His advice was to not try to force the issue but to wait for tempers to calm down. Through an unfortunate turn of events, however, exactly the opposite happened.

Ambassador Niccolini had to meet with the Pope to discuss another topic, and that meeting deteriorated to the point of disastrous consequences, as the frustrated Niccolini described later: "While we were discussing those delicate subjects of the Holy Office, His Holiness exploded into great anger, and suddenly he told me that even our Galilei had dared entering where he should not have, into the most serious and dangerous subject which could be stirred up at this time." Taken aback by this outburst—and not being aware of the fact that by that time Ciampoli had fallen out of favor with the Pope—Niccolini made the additional mistake of trying to argue that the *Dialogo* had been published with Riccardi's and Ciampoli's approval. This struck a nerve with Pope Urban VIII, who now shouted in rage that "he had been deceived by Galileo and Ciampoli," and that "in particular, Ciampoli had dared tell him that Mr. Galilei was ready to do all His Holiness ordered and that everything was fine." All of Niccolini's attempts to convince the Pontiff to give Galileo a chance to explain his actions fell on deaf ears. The Pope screamed angrily that "such is not the custom," and that "he [Galileo] knows very well where the difficulties lay, if he wants to know them, since we [using here the royal *we*] have discussed them with him, and he has heard them from ourselves."

Niccolini's further efforts to at least soften the blow hit a brick wall. The Pope revealed that he had appointed a commission to examine the book "word for word, since one is dealing with the most perverse subject one could ever come across." Finally, repeating his complaint that he had been deceived by Galileo, Urban VIII added that he had, in fact, done Galileo a favor by appointing a special commission and not sending the case directly to the Inquisition.

Faced with this fiasco, Niccolini decided, also on Riccardi's advice, to refrain from any additional steps to try appeasing the Pope, and to solely rely on Riccardi's own endeavors to introduce some corrections into the book, to make it more palatable. Riccardi appeared to be less concerned about the composition of the commission (of which he was a member), estimating that at least two others would treat Galileo with fairness. In this assumption he was wrong, since the Jesuit Melchior Inchofer (most probably a commission member) was a devout anti-Copernican.

At any rate, this was the situation at the beginning of September 1632, and although things looked far from encouraging, there were at least some grounds for mild optimism. However, that was when a new piece of information hit like a bomb.

A GHOST OF THE PAST

As you may recall, sixteen years earlier, Galileo had been summoned to Roberto Bellarmino's palace, where, after a warning by Bellarmino, the Commissary General of the Holy Office, Michelangelo Seghizzi, unnecessarily issued a harsher injunction forbidding Galileo from holding or defending the Copernican system in any publication or in teachings. The document recording this injunction—and Galileo's submission to it—was somehow recovered from the Holy Office's archives around mid-September, and it was brought to the attention of the Pope's commission. With this new evidence at hand, all hopes that the entire affair would be solved by a few corrections to the *Dialogo* were rapidly beginning to fade. In fact, at a meeting of the Congregation of the Holy Office on September 23, it was reported that Galileo had been "deceitfully silent about the command laid upon him by the Holy Office in the year 1616, which was as follows: to relinquish altogether the said opinion that the Sun is the center of the world and immovable and that the Earth moves, nor henceforth to hold, teach, or defend it in any way whatsoever, verbally or in writing, otherwise proceedings would be taken against him by the Holy Office, which injunction the said Galileo acquiesced in and promised to obey." Seghizzi's hand was now reaching for Galileo's neck even from the grave. He'd died in 1625.

In view of this new evidence, the Pope's reaction was swift. He sent a message instructing the inquisitor of Florence to order Galileo to travel to Rome for the whole month of October and appear before the Commissary General of the Holy Office for questioning. Shocked to receive this jarring command, Galileo understood that he had to at least formally demonstrate obedience. At the same time, he was determined to do everything in his (or his friends') power to dodge having to go to Rome, knowing only too well that nothing good could come out of such a trip. As part of these efforts and delay tactics—but also due to his genuinely poor health—he sent a letter on October 13 to Cardinal Francesco Barberini in which he complained that the fruits of his studies and labor "are turned into serious accusations against my reputation," which, as he described, had caused him countless sleepless nights. Galileo made an emotional appeal to Barberini that the Church allow him to either send detailed explanations of all his writings or to appear before the inquisitor and his staff in Florence rather than in Rome.

Ambassador Niccolini, who was supposed to deliver Galileo's letter to the cardinal, hesitated at first, for fear that the letter might hurt more than it could help, but after consulting with Galileo's former student Castelli, he eventually passed it on. At the same time, Niccolini and Castelli sat down with various church officials—Niccolini even met with Pope Urban VIII himself—in a desperate attempt to save the sixty-eight-year-old from having to travel to Rome. All interventions on Galileo's behalf, however, were unsuccessful. The Pope insisted that both Ciampoli and Riccardi "had behaved badly" and had deceived him with regard to the *Dialogo*. In retaliation, on November 23 Ciampoli was effectively exiled from Rome to become governor of a small town. He never returned to Rome.

As for Galileo, the Pontiff instructed the Florentine inquisitor to absolutely oblige Galileo to come to Rome, even though Galileo was given a postponement of a month for the trip. As a last-ditch effort, on December 17 Galileo sent a medical report, prepared by three doctors, which stated that travel would worsen his already grave condition. At that point, the impatient Urban VIII was not prepared to make any further concessions. With the clear intention of intimidating Galileo, he suggested that he would send his own physicians to examine the astronomer—at Galileo's expense!—and that if he were found to be fit for travel, he would be sent "imprisoned and in chains." Confronted with this threat, even the grand duke and his secretary of state informed Galileo that "since it is proper in the end to obey the higher tribunals, it displeases His Highness [the duke] that he cannot bring it about that you would not go." The most that the grand duke could offer at this stage was to help with the travel arrangements and to organize accommodations in Rome at the ambassador's house.

Realizing that he had run out of options, and seriously concerned about what the trip to Rome might entail, Galileo wrote a will. He named his son Vincenzo as heir. He also wrote to his friend in Paris, Elia Diodati, who helped publish Galileo's work outside Italy, saying, "I am sure it [the *Dialogo*] will be prohibited, despite the fact that to obtain the license I went personally to Rome and delivered it into the hands of the master of the Sacred Palace."

Galileo departed for Rome on January 20, 1633, but because of the raging plague, he had to be quarantined before crossing from Tuscany into the territories known as the Papal States, a stop that turned out to be painfully long and unpleasant. Consequently, he arrived in Rome only on February 12, fortunately to the comfort and warm hospitality in the home of Ambassador Niccolini and his wife. After meeting with a few church officials for advice in the first few days, in the following weeks, Galileo barely left the house, since Cardinal Francesco Barberini advised him against socializing, for fear that it "could cause harm and prejudice."

As time went by with very little, clear, detectable action, or any form of communication, Galileo's hopes for a relative benign and peaceful resolution were somewhat resurrected. He also felt encouraged by the fact that he had been allowed to stay at the Tuscan ambassador's house rather than at the quarters of the Holy Office. In his naïveté, Galileo did not understand that after having gone through all the trouble of bringing him to Rome, the Church could not afford to let the affair dissolve. Niccolini's endeavors to achieve a prompt decision by meeting again with the Pope also went nowhere. The Pontiff repeated his stance that "May God forgive Signor Galilei for having meddled with these subjects," since, as Urban VIII kept stating, "God is omnipotent and can do anything; but if He is omnipotent, why do we want to bind him?" The Pope's uncompromising view continued to be that no theoretical understanding of the universe was ever possible.

The state of uncertainty and anxiety continued for about two months. At the beginning of April, however, Galileo was summoned to the Holy Office, and he appeared before the Commissary General on April 12. The only good news that Niccolini could report to the Florentine secretary of state was that Galileo was lodged in the chambers of the prosecutor rather than in the cells usually given to criminals. The prosecutor also allowed for Galileo's domestic to serve him, and food was brought to him from the Tuscan embassy.

The stage was thus set for one of the most famous—or, rather, infamous—trials in history.

CHAPTER 12

The Trial

Galileo's trial began on April 12 and ended on June 22, 1633. The actual interrogations took place during three sessions, on April 12, April 30, and May 10. The Pope's decision was obtained on June 16, and the sentence was delivered six days later. Even though the key charges concerned disobedience to the Church's orders, no other single event represented as clearly the clash between scientific reasoning and religious authority, and its reverberations are felt even today.

The Inquisition, or, more formally, the Congregation of the Holy Office, was composed of ten cardinals appointed by the Pope. The person in charge of the interrogations was the Commissary General, Cardinal Vincenzo Maculano (who was also an engineer), aided by the in-court prosecutor, Carlo Sinceri. As we shall see, while we have a fairly detailed description of what happened inside the courtroom during each session, unfortunately, we have no access to potentially crucial behind-the-scenes information.

SESSION 1: THE SHADOWS OF 1616

After a few preliminary questions, in the answers to which Galileo acknowledged that he assumed he had been summoned to the Holy

Office because of his latest book, the *Dialogo*, the prosecutor moved swiftly to what he regarded as his trump card. With a series of questions, Maculano focused all attention on the injunction of 1616—the document that had been discovered in the archives a few months earlier.

Since this document played a dramatic role in the trial, it is worth recalling the sequence of events that had led to its creation. At a meeting of the Inquisition on February 25, 1616, Pope Paul V ordered Cardinal Bellarmino to summon Galileo and to warn him that he had to abandon the Copernican doctrine. *Only in the event that Galileo refused*, Commissary General Michelangelo Seghizzi was to issue a formal injunction forbidding Galileo from defending, discussing, or teaching Copernicanism in any way. If the astronomer objected even to the injunction, the order was to arrest and prosecute him. At the Inquisition's meeting on March 3, Bellarmino reported that Galileo acquiesced already when issued the warning to stop supporting Copernicanism.

The new document presented at the trial, which was dated February 26, 1616, described a course of events that was different in one important aspect. It stated that *immediately after* Bellarmino's warning, Seghizzi intervened and ordered Galileo to abandon Copernicanism and to not hold, defend, or teach it in any fashion whatsoever, and that Galileo promised to obey. This document seemed to describe a rather premature intervention by the commissary general, perhaps prompted by a mere brief hesitation on Galileo's part after hearing Bellarmino's warning. The document was also in conflict with both Bellarmino's own report to the Inquisition and with Bellarmino's letter to Galileo, issued on May 26, 1616. These discrepancies have spawned an entire series of conspiracy theories suggesting that the injunction document may have been forged, either in 1616 or in 1632. However, a calligraphic analysis of the document obtained in 2009 confirmed that Andrea Pettini, the notary of the Holy Office, recorded all the documents of 1616, thus refuting any suggestions of forgery on the eve of the trial.

When asked specifically about what had been communicated to him in February 1616, Galileo answered without hesitation: "In the month of February 1616, Lord Cardinal Bellarmino told me that since Copernicus's opinion, taken absolutely, was contrary to Holy Scripture, it could neither be held nor defended, but it could be taken and used *suppositionally* [emphasis added]. In conformity with this, I keep a certificate by Lord Cardinal Bellarmino himself, dated 26 May 1616, in which he says that Copernicus's opinion cannot be held or defended, being against Holy Scripture. I present a copy of this certificate, and here it is." At that point, Galileo produced a copy of Bellarmino's letter, which Maculano had no idea existed. This clearly had the potential of being a critical moment from a legal perspective, since while the injunction issued by Commissary General Seghizzi (with Bellarmino present) spoke of not to "hold, teach, or defend in any way, either verbally or in writing," Bellarmino's letter used the much milder language "not to hold or defend Copernicanism." Evidently caught by surprise, Maculano tried to press Galileo on whether there were any others present at that meeting, and Galileo answered that there were some Dominican fathers present whom he did not know, nor had he seen since. Still insisting, Maculano asked Galileo specifically about the wording of the injunction. Galileo's response, which sounds sincere, was not phrased in the most advantageous way to aid his defense:

I do not recall that such injunction was given to me any other way than orally by Lord Cardinal Bellarmino. I do remember that the injunction was that I could not hold or defend, or maybe even that I could not teach. I do not recall, further, that there was a phrase "in any way whatever," but maybe there was; in fact, I did not think about it or keep it in mind, having received a few months thereafter Lord Cardinal Bellarmino's certificate dated 26 May, which I have presented and in which is explained the order given to me not to hold or defend the said opinion. Regarding the other two phrases in the said injunction now mentioned, namely

not to teach and in any way whatever, I did not retain them in my
memory, I think because they are not contained in the said certificate, which I relied upon and kept as a reminder.

Unfortunately, by allowing for the possibility that the injunction
might have been more restrictive than Bellarmino's letter, Galileo inadvertently weakened the protection that Bellarmino's softer formulation offered him. Without this apparently sincere, albeit tentative
admission, there would have been the ambiguous legal issue of the
two documents—Bellarmino's letter, on one hand, and the injunction
document, on the other—being incompatible with each other. It is
difficult to know why Galileo chose to acknowledge tentatively something that happened so many years earlier and that he legitimately
didn't recall precisely. It could be that he mistakenly thought it wasn't
that important, especially given the line of defense that he was about
to adopt. Indeed, the next set of queries concerned the imprimatur—
the permission to write and publish the *Dialogo*.

The first question was perhaps the most problematic. Galileo was
asked whether he had requested permission to write the book. The
simple answer was, of course, no. However, acknowledging that fact
without any explanation, combined with the prevailing perception
that the book advocated Copernicanism, would have been tantamount to an immediate admission of guilt. Galileo therefore decided
to rely on the fact that the added preface in particular and the final
summary of the book rendered his opinion on Copernicanism inconclusive and his support for it neither explicit nor absolute. Accordingly, he claimed that he had not felt that he needed permission, since
his goal had not been to support Copernicanism but to refute it. Any
lawyer today would have told Galileo that this choice of word (*refute*)
was not particularly credible, given the actual contents of the *Dialogo*,
and Galileo's statement must have raised a few eyebrows in the court.

Why did Galileo make such a claim? It is difficult to judge what
was in the mind of an old man fearing imprisonment. Galileo was
probably attempting to give more weight to his statement in the

preface, where he seemingly expressed his support for the decree against Copernicanism of 1616. It is also possible that in fact he used a weaker expression and "refute" was inserted by the church officials recording the proceedings, in their attempt to portray Galileo as deceitful and manipulative.

Rather than seeking to contradict Galileo, Maculano moved on to the next question, on whether Galileo had sought permission to print the book. On this, Galileo had an answer that on the face of it appeared convincing: he had not just one imprimatur but two. One was from the Master of the Sacred Palace, Niccolò Riccardi in Rome, and the second from the Inquisitor of Florence, Clemente Egidi. This, in principle, could have been an extremely strong point in favor of Galileo. Could the Church really condemn a book that had been approved for publication twice by the very church officials in charge of censorship?

Perfectly understanding the problem the prosecution was facing, Maculano attempted to establish that Galileo had, in fact, been disingenuous in his request for permission. He therefore asked Galileo whether he had revealed to Riccardi the existence of the injunction of 1616. Galileo replied that he had not, arguing again that he had considered such a notification unnecessary, given that his purpose was not to defend Copernicanism but to demonstrate that no world model could be made conclusive—precisely in line with the Pope's views. Here Galileo may have made another tactical mistake. To be consistent with his previous statements, he could have claimed that he had not informed Riccardi of the 1616 injunction simply because he didn't remember it. Galileo's numerous missed opportunities to exploit legal loopholes to strengthen his case does leave us with the impression that his answers may have been sincere, at least from his own perspective, or that they have been misrepresented in the written document.

With this indecisive exchange, the first session came to a close. Following the proceedings, Galileo was asked, as the protocol required, to sign a deposition and then was taken to be detained on the premises of the Dominican convent of Santa Maria Sopra Minerva, the location where the Holy Office held the hearings.

From the point of view of an objective audience, one could argue that the first session of the trial ended in a legal draw. Whereas Maculano surely surprised and scared Galileo with the introduction of Seghizzi's injunction document, Galileo produced his own surprise in the form of Bellarmino's letter. Since the *Dialogo* consisted of a critical examination of arguments both in favor of and against Copernicanism, it could be regarded as violating the prohibition of "teaching" contained in Seghizzi's injunction. At the same time, Galileo could maintain that he had complied fully with Bellarmino's letter, which had merely forbidden explicit support for Copernicanism. With both Bellarmino and Seghizzi dead at the time of the trial, the two conflicting documents created a virtual impasse from which there seemed to be no easy way out. The two permissions to print the book—one supposedly granted by Riccardi in Rome (albeit somewhat problematically, since the book was actually printed in Florence) and the other by Egidi in Florence—further complicated the issue, and must have given Maculano a strong sensation of an impending stalemate.

We may wonder why Egidi and Riccardi gave the imprimatur in the first place, given the contents of the book, which other church officials clearly found objectionable. One can only speculate that knowing the rather intimate friendship between Urban VIII and Galileo up to around 1630, they both must have assumed that the book had fully met with the Pope's at least implicit approval, especially since the Pontiff's views had been explicitly included (though coming from Simplicio's mouth). Alas, by 1633, the entire personal and political scenes had changed. Egidi's decision was almost certainly influenced also by the fact that Galileo had always been one of the Grand Duke's favorites.

Disastrously for Galileo, the three members of the new special commission appointed to carefully examine the *Dialogo* in order to determine whether Galileo held, taught, or defended in any way the propositions that the Sun is at rest and the Earth is moving, issued their individual reports on April 17. They all concluded definitively that the book violated Seghizzi's 1616 injunction, even though two of them did not explicitly pronounce that Galileo *held* the condemned Copernicanism.

The report by Jesuit Melchior Inchofer, who was a strong opponent of Copernicanism and a supporter of Christoph Scheiner, was particularly long, extremely detailed, and devastatingly damning. It started with a grave indictment: "I am of the opinion that Galileo not only teaches and defends the immobility or rest of the sun or center of the universe, around which both the planets and the earth revolve with their own motions, but also that he is vehemently suspected of firmly adhering to this opinion, and indeed that *he holds it*" [emphasis added]. Inchofer also surmised that one of Galileo's aims was to specifically attack Scheiner, who had written against Copernicanism. As expected, the Congregation of the Index promptly approved the special commission's reports on April 21.

A letter discovered only in 1998 (and published in 2001) in the archives of the Congregation for the Doctrine of the Faith, gave rise to speculation about what happened next in the trial. The letter, written by Maculano and addressed to Cardinal Francesco Barberini, was dated April 22, just one day after the Congregation approved the judgment against the *Dialogo*. Maculano rather compassionately described the situation:

> Last night, Galileo was afflicted with pains which assaulted him, and he cried out again this morning. I have visited him twice, and he has received more medicine. This makes me think that his case should be expedited very quickly, and I truly think that this should happen in light of the grave condition of this man. Already yesterday the Congregation decided on his book, and it was determined that in it he defends and teaches the opinion which is rejected and condemned by the Church, and that the author also makes himself suspected of holding it. That being so, the case

could immediately be brought to a prompt settlement, which I
expect is your feeling in obedience to the Pope.

In other words, Maculano started to think of ways to finish the trial
as quickly as possible, perceiving that a certain level of guilt had already been established. In the seventeenth century, just as today, a
clear and simple method to shorten legal processes was through a
plea bargain.

One Galileo scholar therefore suggested that this was precisely
what Maculano was trying to achieve: Galileo would plead guilty
to some relatively minor offense, such as being "rash," in writing the
Dialogo, and the prosecution would impose a lighter penalty. A letter
written by Maculano to Cardinal Francesco Barberini on April 28
seems to support this interpretation. In it, Maculano first describes
how he had been successful in convincing the cardinals in the Congregation of the Holy Office "to deal extrajudicially with Galileo."
He then adds that in a meeting he had with Galileo, the latter also
"clearly recognized that he had erred and gone too far in his book,"
and that "he was ready for a judicial confession." Maculano concludes
by saying he believes that in this way the trial could be "settled without difficulty," with the court maintaining its reputation and with
Galileo knowing that a favor was done to him. Everything appeared
set, therefore, for a rapid and relatively benign end to the trial, with
a ruling stating that, in Maculano's words, "he [Galileo] could be
granted imprisonment in his own house, as Your Eminence [Cardinal
Barberini] mentioned."

SESSIONS 2 AND 3: A PLEA BARGAIN?

If Maculano and Galileo had indeed struck some sort of deal, then at
least the format of the next two sessions would have been prescribed:
Galileo should have confessed in session 2, and then he would have
been allowed to present a defense in the following session. The trial
indeed seemed to proceed along these lines. In the second session,

Galileo asked for permission to make a statement, and when this was granted, he explained that he had spent the time since the first session reviewing the *Dialogo* to check whether he had inadvertently disobeyed the injunction of 1616. Through this scrutiny, he said, he discovered "several places to be written in such a way that a reader, not aware of my intention, would have had reason to form the opinion that the arguments for the false side [Copernicanism], which I intended to confute, were so stated as to be capable of convincing, because of their strength." Galileo was again repeating here his questionable claim that his intention in the *Dialogo* was to refute Copernicanism. Given that he had time to reflect on the matter, it could be that he was deliberately using the same language he had employed in the first session, to make his confession more credible. "My error then was, and I confess it, one of vain ambition, pure ignorance, and inadvertence," he added. Finally, and sadly, Galileo even proposed that he could extend by a day or two the discussion in the *Dialogo* in a new book, to clarify the falsity of the Copernican view. The court ignored this suggestion.

Galileo's preposterous overture could, it seems, be justified only in two ways. Either in spite of Maculano's generally friendly demeanor, the astromer still had a mortal fear of being tortured, or he thought that in this way he could perhaps still salvage the *Dialogo* from being condemned. Either way, Galileo's reaction was a clear demonstration of what intimidation could cause even to the most independent of thinkers, thus evoking horrifying memories of totalitarian regimes past and present. Cases such as those of self-exiled dissident Saudi Arabian journalist Jamal Khashoggi and Russian defector Alexander Litvinenko, both murdered by their native countries' governments, immediately come to mind. Following the second session, Galileo was allowed, after signing the deposition, to return to the Tuscan ambassador's house, "having considered the bad health and advanced age."

Maculano's plan to end the trial rapidly and relatively benevolently is further suggested by a note from Ambassador Niccolini dated May 1, in which he wrote: "Father Commissionary himself [Maculano] also manifests the intention of wishing to arrange it that this cause be dropped and that silence be imposed on it; and if this is achieved, it will shorten everything and will free many from troubles and dangers."

Whether or not a plea bargain had indeed been reached, the third session was certainly consistent with such an agreement. Galileo presented the original of Bellarmino's letter, as well as a personal statement of defense in which he explained that he had used this document as his only guide. Consequently, he clarified, he felt that he was "quite reasonably excused" from having to inform Father Riccardi, and if he had violated the more stringent restrictions imposed by the injunction, which he had totally forgotten, this was "not introduced through the cunning of an insincere intention, but rather through the vain ambition and satisfaction of appearing clever above and beyond the average among popular writers." In conclusion, Galileo expressed his willingness to make any amends ordered by the court, and he asked for leniency, based on his age and infirmity. This last request raises doubts about a plea bargain having been accepted, for if it had indeed been in place, presumably the penalty had also been discussed already. It may be, however, that this was a formality required to further justify a less severe punishment.

The only procedural step left to be carried out was for a legal summary of the proceedings to be written and delivered to the Inquisition and the Pope. Ambassador Niccolini had an audience with Urban VIII on May 21, and was assured by him and by Cardinal Francesco Barberini that a painless conclusion of the trial was imminent. The fact that Galileo was allowed to leave the house for short walks at this stage also hinted at a sympathetic resolution. Galileo himself wrote an optimistic letter to his daughter, Sister Maria Celeste, thanking her for her prayers on his behalf.

CHAPTER 13

I Abjure, Curse, and Detest

To assist the inquisitors in reaching a verdict, protocol required that the assessor in the staff of the Holy Office, Pietro Paolo Febei, write a summary of the trial proceedings. That internal document, which was inaccessible to Galileo, was to be distributed only to the Congregation and the Pope.

As it turned out, the summary revealed a clear intention to present Galileo in the worst possible light. It included misleading, irrelevant, and even some downright false material that could be perceived as incriminating, while it deliberately omitted details that could have helped Galileo's case.

Instead of dealing directly with the *Dialogo*, the summary started with a synopsis of the old complaints made against Galileo in 1615 by the Dominican Niccolò Lorini and the preacher Tommaso Caccini, which had been based largely on vague hearsay. Some of those ridiculous and false accusations alleged such things as Galileo's having been heard to state that God was an accident, or that the miracles performed by saints were not true miracles. Even references to the famous *Letter to Benedettó Castelli* and *Letter to the Grand Duchess Christina* found their way into the summary, without mentioning the fact that *Letter to Castelli* (or at least the slightly moderated version of it) had been vetted and judged to be inoffensive and the case dismissed. There is no question that the inclusion of all these old files and others was done with the aim of enhancing the impression of a serial offender. To this end, the summary also contained Caccini's fallacious claim that in his *Letters on Sunspots*, Galileo had explicitly defended Copernicanism. While Galileo undoubtedly thought that the observations of sunspots supported the Copernican model, his book never made any categorical statements. Even in the description of Bellarmino's warning and Seghizzi's injunction, the summary contained small but important inaccuracies. In particular, the summary did not mention at all the fact that this was supposed to be a two-step process, and that, in fact, the stricter injunction was unjustified. Crucially, the summary played down the fact that Bellarmino's letter did not contain the additional restrictions of "not to teach" and "in any way," saying that it was Bellarmino himself, not Seghizzi, who issued the more specific injunction. In this way, rather than conveying the correct fact that Bellarmino's letter and Seghizzi's injunction were in conflict with each other, the summary gave the impression that they were complementary.

Why was the summary so biased against Galileo? And perhaps even more intriguingly, if indeed there was a plea bargain, what had happened to it? We shall probably never know the precise answers to these questions. The summary itself was most probably written by the assessor in the Holy Office, Febei, who was perhaps aided by Cardinal Maculano's in-court interrogator, Carlo Sinceri.

Why would these two, perhaps with the help of others, write such a flawed, unfair, and damning summary report of the trial? We can only speculate. Presumably, among the cardinals in the Congregation and among the officers of the court, there were some—perhaps even a majority—who disagreed with the attempt to reach a rapid conclusion and a reduced sentence. This "severe" group may have included the Pope himself, and certainly Inchofer. After all, the special commission that had examined the *Dialogo* did conclude unanimously that, in this book, Galileo disobeyed the injunction of 1616. Galileo's statement that he tried to refute Copernicanism in the book could not be taken seriously by anyone who had read it or at least read Melchior Inchofer's report. Consequently, these severe cardinals, who were perhaps less inclined to be forgiving with Galileo from the start, would have voted against any plea-bargain-like endeavor and for a more rigorous punishment—especially after having read the summary. The more hard-line cardinals may have also wanted, for political reasons, to keep the Pope as distant as possible (as perceived by all Catholics) from the Galileo scandal, while having to justify a condemnation of the celebrated scientist.

While the Pope almost certainly had made up his mind long before and wasn't involved in the details of the trial itself, once the summary was brought to his attention, surely any hope for a benign ending was doomed. Based on the Pontiff's earlier complaints about Galileo to Ambassador Niccolini, it seems quite plausible that Urban VIII had not fully recovered from the feeling that in writing the *Dialogo*, Galileo had deceived and betrayed him—even though he claimed otherwise, in a later conversation with the French ambassador—and he sought revenge. When coupled perhaps to his political sense that he needed at that time to demonstrate toughness on religious matters and that Galileo's book was in his words, "pernicious to Christianity," the Pope was probably pleased with having been given the opportunity to impose serious punitive consequences on Galileo. One can even speculate that the summary could not have been as harsh as it was without its authors being aware of the Pope's insinuated approval.

In the first English biography of Galileo, Thomas Salusbury's *Life of Galileo* (only one copy of which survived the Great Fire of London in 1666), Salusbury, a Welsh writer who lived in London in the mid-seventeenth century, advanced an original thesis even for the root cause of Galileo's trial itself. According to Salusbury, a personal reason combined with a political backdrop prompted the Pope to put Galileo on trial. The personal motive was supposedly Pope Urban's uncontrolled rage about being caricatured as Simplicio. While this is not altogether implausible, there is no documented evidence for this being the source of the original charge against the book, and, in fact, this particular accusation first surfaced only in 1635, more than two years after the trial. There is no question that Galileo never intended to insult the Pope in this way, and even after the rumor started circulating, the French ambassador and Castelli managed to convince the Pontiff that there was no truth in it. The political hypothesis was even more intriguing. Salusbury wrote:

"Add to this, that he [the Pope] and his fastidious Nephews, Cardinal Antonio and Cardinal Francisco Barberini (who had embroyled all Italy in Civil Wars by their mis-government) thought to revenge themselves upon their Natural Lord and Prince, the Great Duke, by the oblique blows which they aimed at him through the sides of his Favourite."

In other words, Salusbury suggested that Galileo's trial represented a papal retribution to Galileo's patrons, the Medicis, for their rather lukewarm military support in the Thirty Years' War.

Either way, on June 16 the Holy Office met and issued a hardhearted decision:

Sanctissimus [the Pope] "decreed that the said Galileo is to be interrogated on his intention, even with the threat of torture; and this having been done he is to abjure under vehement suspicion of heresy in a plenary session of the Congregation of the Holy Office; then is to be condemned to the imprisonment at the pleasure of the Holy Congregation, and ordered not to treat further, in whatever manner, either in words or in writing, on the mobility of the Earth and the stability of the Sun or against it; otherwise he will incur the penalties of relapse.

In addition, the Pope decided that the book entitled *Dialogo di Galileo Galilei Linceo* would be placed in the *Index of Forbidden Books*. The Church also took steps to make this decision widely known, both to the public and to other mathematicians. While torture would almost certainly not have been applied to a person of Galileo's age, the mere formal threat with torture must have been horrifying for him.

On June 21 Galileo was summoned for the official interrogation about his "intentions," to establish whether he had committed his crimes innocently or deliberately. As part of this ritual, he was asked specifically—in three different ways—whether he believed in the Copernican model. The by now broken and defeated old man answered that following the decree of 1616, he concluded that the Ptolemaic, geocentric scenario was the correct one. We can only imagine how much it must have pained Galileo to utter these words. He further insisted that in the *Dialogo*, his goal was only to demonstrate that on the basis of science alone, no conclusive opinion could be reached, and one therefore had to rely on the "determination of more subtle doctrines." In other words, the Church's opinion.

What happened on the following day remains one of the most shameful events in our intellectual history. In front of the inquisitors, Galileo, on his knees, was informed that he had rendered himself "vehemently suspected of heresy, namely, of having held and believed a doctrine which is false and contrary to the divine and Holy Scripture: that the sun is the center of the world and does not move from east to west, and the earth moves and is not the center of the world, and that one may hold and defend as probable an opinion after it has been declared and defined contrary to the Holy Scripture."

The cardinals of the Holy Office then added, as if mercifully:

"We are willing to absolve you from them [all the censures and penalties] provided that first, with sincere heart and unfeigned faith, in front of us you abjure, curse, and detest the above-mentioned errors and heresies, and every other error and heresy contrary to the Catholic and Apostolic Church, in the manner and form we will prescribe to you."

The verdict included "formal imprisonment" at the pleasure of the Holy Office; having to recite seven penitential Psalms once a week for three years; and the *Dialogo*'s being banned.

We do not know if Cardinal Francesco Barberini's (and two others') absence at the signing of the sentence reflected their disapproval of the condemnation or was merely the result of a scheduling conflict. We do know that at the time of the abjuration itself, Francesco Barberini had a meeting with Pope Urban VIII.

Again on his knees, Galileo read the text of the abjuration given to him:

> I, Galileo, son of the late Vincenzo Galilei of Florence, seventy years of age, arraigned personally for judgment, kneeling before you Most Eminent and Most Reverend Cardinals Inquisitors-General against heretical depravity in all Christendom, having before my eyes and touching with my hands the Holy Gospels, swear that I have always believed, I believe now, and with God's help I will believe in the future all that the Holy and Apostolic Church holds, preaches, and teaches.

Then, after committing "to abandon completely the false opinion" of Copernicanism, Galileo read the essence of the abjuration:

> Therefore, desiring to remove from the minds of Your Eminences and every faithful Christian this vehement suspicion, rightly conceived against me, with a sincere heart and unfeigned faith, I abjure, curse, and detest the above-mentioned errors and heresies, and in general each and every other error, heresy, and sect contrary to the Holy Church, and I swear that in the future I will never again say or assert, orally or in writing, anything which might cause a similar suspicion about me; on the contrary, if I should come to know any heretic or anyone suspected of heresy, I will denounce him to this Holy Office, or to the Inquisitor or Ordinary of the place where I happen to be."

The humiliation associated with having to utter these words, which undermined much of his life's work, must have been unimaginable. Those historians of science who attempt to argue that had Galileo been generally less combative, things would have ended better, ignore the simple fact that he was forced to abjure his deep convictions under the threat of torture. Galileo's judges could not know that over the following four centuries, this denigrating event would be transformed into one of the most deplorable acts of the Inquisition.

Legend has it that upon leaving the scene, Galileo mumbled, "E pur si muove"—"And yet it moves"—referring to the Earth. The earliest source of this story was claimed to be a painting from the mid-seventeenth century (1643 or 1645). In this painting, Galileo is depicted in prison, looking at a diagram of the Earth orbiting the Sun that he had scratched on the wall, with those words underneath it. Assuming that the painting was indeed from 1643 or 1645, this was taken as proof that the legend started circulating very shortly after Galileo's death. A thorough investigation I conducted in 2019 raised some serious doubts about the authenticity of this painting.

The first appearance of the legendary motto in a printed book was in the eighteenth century, in *The Italian Library* by an Italian living in London, Giuseppe Baretti. Galileo could not have muttered these words in front of the inquisitors, but it is not impossible that he did utter some version of this phrase, which was surely on his mind, to one of his friends. At any rate, Galileo's bitterness about the trial and his contempt for the inquisitors continued to occupy his mind for the rest of his life.

Today the phrase "And yet it moves" has become a symbol of intellectual defiance, implying that "in spite of what you believe, these are the facts." Unfortunately, in an era of "alternative facts," there appear to be more and more occasions where the use of the phrase is appropriate.

Did the Church act within its legal authority in terms of the charges brought against Galileo? From its very narrow point of view, most probably yes, given the warning of Bellarmino and the injunction issued to Galileo by Seghizzi. Galileo was convicted essentially because of two facts: first, for having violated the 1616 injunction, and second, for having obtained the imprimatur to print the *Dialogo* "artfully and cunningly" by not revealing the injunction to Riccardi and Egidi. In this sense, the conviction was justified. The abjuration was also a necessary step, since without it, the "suspicion of heresy" would have turned into actual heresy, for which, as we know, Giordano Bruno had been burned at the stake.

Judging the affair from a broader perspective, however, there is a more important point to be made. Putting Galileo on trial, imprisoning him, and forbidding his book, was wrong not just because Galileo was right about the science of the solar system. These actions against intellectual freedom and, by implication, even against religious beliefs, would have been wrong even if the geocentric model were the correct one. The much bigger lesson from the Galileo affair is that no officialdom, be it religious or governmental, should have the authority to impose punishments on scientific, religious, or any other type of opinions (whether correct or incorrect), as long as those neither harm, nor incite others to harm, anybody else. This is precisely why the real Galileo affair was planted in humanity's conscience *after* the verdict and abjuration in Galileo's trial. In that further-reaching affair, the inquisitors became the culprits, and the affair itself remains a constant reminder that the freedom to express truths should never be taken for granted.

CHAPTER 14

One Old Man, Two New Sciences

Galileo's sentence included detention. He therefore had to be informed where that would take place. Fortunately, the Pope commuted his sentence to house arrest and, on June 30, 1633, allowed him to start the imprisonment at the house of Ascanio Piccolomini, the archbishop of Siena, where Galileo spent about a half year. In spite of the restriction on his personal freedom, Galileo enjoyed his stay in the house of the receptive and learned archbishop, who regarded Galileo as "the greatest man in the world." It was in Piccolomini's house where Galileo started to work on his last great book, *Discorsi*, which summarized all of his experimental work in Padua and his insights in mechanics. With an ironic historical twist, Galileo's mechanics were precisely the tools that Sir Isaac Newton later needed to prove Copernicanism correct.

Throughout his time in Siena, however, what Galileo really wanted more than anything was to return to his home in Arcetri, near Florence. In his absence, the home was managed by his daughter, Sister Maria Celeste, from her convent nearby. From her wonderful letters, Galileo learned that lemons, beans, and lettuce were thriving, and that the wine from his casks tasted good. This young woman was able to comfort her old father even at his darkest hours with her remarkable yet affectionate calm. Having given up other forms of love, she lavished her most tender love on Galileo and wrote to him following his trial:

> As much as the news of Your Honor's new affliction was sudden and unexpected, that much more my soul was pierced with extreme grief in hearing the decision finally taken both about your book and Your Honor's person . . . now is the time to avail yourself more than ever of that prudence which the Lord God has granted you, bearing these blows with that strength of spirit which your religion, your profession, and your age demand.

In December 1633 the Pope finally allowed Galileo to return to Arcetri, where he was to be in perpetual house arrest and strictly forbidden from turning the place into a gathering center for intellectuals, scientists, and mathematicians. While Galileo was very happy to be back home and close to his loving daughter, this happiness was rather short-lived. Sister Maria Celeste died at age thirty-three only three months after Galileo's return. Galileo was devastated. "I had two daughters whom I much loved, especially the elder, a woman of fine mind and singular goodness and most affectionate to me," he wrote to his friend Elia Diodati in Paris.

Seeking comfort in his work, Galileo managed to complete *Discorsi* in 1635, and his original intention was to publish the book in Venice. This turned out to be easier said than done. All the local inquisitors in Italy had received from the Roman Inquisition the full text of Galileo's sentence and abjuration. The inquisitor of Venice informed Galileo's friend (and Paolo Sarpi's biographer) Fulgenzio Micanzio, that Rome had issued an order prohibiting the publication of *any* book by Galileo, including the reprinting of previously published books. Consequently, Galileo clandestinely sent copies of *Discorsi* to friends outside of Italy with the hope of securing a publisher somewhere outside the circle of dominance of the Catholic Church and the Jesuits.

One of those friends, military engineer Giovanni Pieroni, tried unsuccessfully to publish the book in Prague. He expressed his frustration in a letter to Galileo: "What an unhappy place we live in," he complained, "where there reigns a determined resolution to exterminate all novelties, especially in science, as if we already knew everything knowable." Indeed, what eventually marked the scientific revolution, in which Galileo played a major role, was the recognition that humans did not know everything, and that exploration, observation, and experimentation offered the best way to acquire new insights and knowledge. Eventually Louis Elsevier, an able publisher in the Protestant university city of Leiden, the Netherlands, published the book in 1638. Elsevier managed to obtain one copy of *Discorsi* when he visited Venice. A second copy was smuggled to him by the French ambassador to Rome, a loyal admirer of Galileo's who received permission to stop at Galileo's home upon his return to France.

DISCORSI

Discorsi marked the final chapter of Galileo's scientific story (Figure 14.1 shows the title page). Like the *Dialogo*, it again involved the cast of Salviati, Sagredo, and Simplicio, this time discussing topics in mechanics rather than lofty world systems. The "two sciences" mentioned in the title referred to a mathematical description of the nature of matter and of material strength, and to the topic of the principles of motion.

The book included Galileo's important discoveries in mechanics, such as the fact that in the absence of air resistance heavy and light bodies fall at the same rate (rather than heavier bodies falling faster, as Aristotle had pronounced). To prove this point, Galileo used a beautiful "thought experiment." Imagine, he said, that you join together a light body and a heavy one. According to Aristotle, since the lighter body falls slower, it should slow down the heavier body, and

Figure 14.1. Title page of *Discorsi*.

the combined body should fall *slower* than the heavy body by itself. On the other hand, Galileo pointed out, one could consider the two joined bodies as one body that is even heavier than the original heavy body; therefore, according to Aristotle, they should fall *faster* than the individual heavy body—a clear contradiction.

To justify his Paduan experimental results with balls rolling down inclined planes as opposed to truly free-falling, Galileo had to show how the motion of such rolling balls is related to a body in free fall. To this goal, he noted that the speed balls reach after rolling down an inclined plane depends only on the *vertical* distance the balls had covered, and not on the angle at which the plane was tilted. In this sense, a free-falling body could be thought of as a ball rolling down a vertical plane.

One of Galileo's crowning achievements was calculating the trajectory traced by projectiles. This came out of an experiment conducted in 1608, which looked something like this: An inclined plane was placed on top of a horizontal tabletop. A ball rolled down the inclined plane, then on the horizontal table, eventually shooting off the table's edge along a trajectory that finally hit the ground. By measuring the horizontal and vertical distances traveled, and by understanding that the horizontal motion (while in the air) is nearly at a uniform speed (since only the air resistance acts to slow it down slightly), while the vertical motion is in free fall, he was able to determine the geometrical shape of the trajectory. Basically, the vertical distance through which the body falls is proportional to the square of the horizontal distance traveled. That is, a ball that travels twice farther horizontally falls four times farther vertically. The trajectory precisely delineates the curve known since antiquity as a parabola.

Overall, the "New" in the title of Galileo's book didn't refer so much to the subjects discussed in the book. After all, people had used beams of wood for construction (and were therefore interested in their strength) thousands of years before Galileo, and bows and catapults shot projectiles into the air in ancient Greece—not to mention the biblical story of David and Goliath. What was new in Galileo's discussion was the way in which mechanics was treated. Through an ingenious combination of experimentation (for example, with inclined planes), abstraction (discovering mathematical laws), and rational generalization (understanding that the same laws apply to all accelerated motions), Galileo established what has since become the modern approach to the study of all natural phenomena.

Perhaps the best demonstration of the evolution in Galileo's thoughts on mechanics was provided by his law of inertia, which later became known as Newton's first law of motion. Starting from Aristotle's notions of "natural" and "violent" motions, and realizing that even fire would move downward if not for the buoyancy provided by the air, Galileo started thinking about how a body would behave if no force whatsoever acted upon it. Finally, in *Discorsi*, he found the answer:

"Along a horizontal plane, the motion is uniform [constant speed], since here it experiences neither acceleration nor retardation." Then came the punchline: "any velocity imparted to a moving body will be rigidly maintained as long as the external causes of acceleration or retardation are removed, a condition which is [experimentally, approximately] found only on horizontal planes." This velocity, he added, "would carry the body at a uniform rate to infinity."

Newton's first law of motion indeed states that an object will remain at rest or in uniform motion along a straight line unless acted upon by an external force. Formulating this law required on Galileo's part imagining a world without friction, which is much harder than it may seem. Friction is such a common feature in all of our everyday experiences—it allows us to walk and to hold objects in our hands, and it slows down every motion we see—that to envisage what would happen without it required a truly phenomenal power of abstraction.

This was Galileo at his best. He established the belief in the existence of what we call today the laws of nature, which are universally valid and perpetually reproducible. Nature cannot be deceitful, or, as Einstein put it several centuries later: "The Lord God is subtle, but malicious he is not." In the introduction to the discussions of the third day of *Discorsi*, Galileo wrote what could be regarded as his own summary of his contributions:

> My purpose is to set forth a very new science dealing with a very ancient subject. . . . I have discovered by experiment some properties of it which are worth knowing and which have not hitherto been either observed or demonstrated . . . what I consider more important, there have been opened up to this vast and most excellent science, of which my work is merely the beginning, ways and means by which other minds more acute than mine will explore its remote corners.

CHAPTER 15

The Final Years

The year 1634 was one of the worst in Galileo's life. In addition to him being under house arrest, not only did his beloved daughter die, but Galileo also had to support the few members of his brother Michelangelo's family who survived the plague in Munich. All that the distraught Galileo could do was to send some money and invite them to come to Arcetri to be together.

Galileo's eyes were also starting to bother him. At first, he attributed his failing vision to the grueling readings he had to put in while preparing the *Dialogo*. Although he continued to work on problems related to navigation at sea—and he even embarked on a series of experiments with pendulums—Galileo was rapidly losing his eyesight, first in his right eye, then in the left. From his descriptions of the progression of his blindness, modern ophthalmologists diagnosed his condition as bilateral uveitis—an inflammation in the middle layer of the eye—or creeping angle closure glaucoma. He was totally blind during the last four years of his life.

Not being able to look through his precious telescope anymore, the distressed Galileo wrote to his friend Diodati:

Alas, my good sir, your dear friend and servant Galileo is irreparably and completely blind, in such a way that the sky, that world and that universe, which with my wondrous observations and clear demonstrations I amplified a hundred and a thousand times over what was believed most commonly by the learned of all past centuries, is for me now so diminished and narrowed that it is no greater than what my body occupies.

It was during that agonizing period that poet John Milton visited him, in 1638. Following the general perception that "travel broadens the mind," Milton was on a European tour in which he tried to meet with as many intellectuals as he could. Having met Galileo's son, Vincenzo, at a literary society meeting in Florence, Milton jumped at the opportunity to be introduced to the most famous scientist in Europe. Not much is known about what transpired at the meeting, but there is no doubt that Galileo's discoveries, his trial, and the con-demnation of his book had a great influence on Milton. In *Paradise Lost*, Milton refers to the "glass of Galileo," and to the innumerable stars discovered by him:

> *Of amplitude almost immense, with stars*
> *Numerous, and every star perhaps a world*
> *Of destined habitation.*

In 1644 Milton published a pamphlet entitled *Areopagitica*—the title inspired by the name of a hill in ancient Greece where the Council of Athens used to meet—in which he argued against censorship of books. This essay is still regarded today as one of the most impas-sioned pleas for freedom of speech, and the US Supreme Court referred to it in interpreting the First Amendment to the US Consti-tution.

In *Areopagitica*, Milton wrote ardently:

And lest some should persuade ye, Lords and Commons, that these arguments of learned men's discouragement at this your order, are mere flourishes, and not real, I could recount what I have seen and heard in other Countries, where this kind of inquisition tyrannizes; when I have sat among their learned men, for that honour I had, and been counted happy to be born in such a place of Philosophic freedom, as they supposed England was, while themselves did nothing but bemoan the servile condition into which learning amongst them was brought; that this was it which had damped the glory of Italian wits; that nothing had been there written now these many years but flattery and fustian [pompous speech]. There it was that I found and visited the famous Galileo grown old, a prisoner to the Inquisition, for thinking in Astronomy otherwise than the Franciscan and Domican licencers thought.

Sadly, Milton diagnosed the situation correctly. For a while, at least, Galileo's fate exerted a chilling, impeding effect on progress in deciphering the cosmos. The great French philosopher René Descartes wrote in November 1633 a letter to his and Galileo's friend the polymath Marin Mersenne, in which he lamented:

I inquired in Leiden and Amsterdam whether Galileo's *World System* was available, for I thought I had heard that it was published in Italy last year. I was told that it had indeed been published, but that all the copies had immediately been burnt at Rome, and that Galileo had been convicted and fined. I was so astonished at this that I almost decided to burn all my papers, or at least to let no one see them.

Mercifully, Galileo eventually prevailed. Already in 1635 a Latin translation of the *Dialogo* was published in Protestant Strasbourg, France. Slowly but surely, the Church itself started changing. In 1757 Pope Benedict XIV, realizing that the Catholic astronomers themselves were using the Copernican scenario, rescinded the prohibition against books that discussed the central tenets of Copernicanism: the Earth's revolution around the Sun and the Sun's immobility. In 1820 the master of the Sacred Palace refused to allow printing for a book that described the heliocentric model, but he was overruled by Pope Pius VII, who decreed that "no obstacles exist for those who sustain Copernicus's affirmation regarding the earth's movement." In 1822 the Church even declared penalties for prohibiting the publication of books that presented the Earth's revolution around the Sun as an established scientific fact. Finally, in 1835 both Copernicus's book and the *Dialogo* were removed from the *Index of Prohibited Books*.

Physically, Galileo was deteriorating rapidly during his last four years. Modern medical researchers have speculated that he suffered from an immune rheumatic disease, reactive arthritis. An inquisitor who was sent to check on whether Galileo's complaints were justified found that he was suffering from severe insomnia and that "he looks more like a cadaver than a living person." Still, even though the Pope allowed Galileo to move to his son's house to be able to receive better medical care, he insisted on prohibiting him from discussing Copernicanism under any circumstances. Galileo contracted fever in November 1641, and he died on the evening of January 8, 1642, presumably of congestive heart failure and pneumonia. His son, Vincenzo, and his students Vincenzo Viviani and Evangelista Torricelli, a talented experimentalist who invented the barometer, were at his side. Figure 8 in the color insert shows Viviani and Galileo. Viviani movingly described Galileo's passing:

"At the age of seventy-seven years, ten months, and twenty days, with philosophical and Christian constancy, he rendered his soul to his Creator, sending it forth, as far as we can believe, to enjoy and admire more closely those eternal and immutable marvels, which that soul, by means of weak devices with such eagerness and impatience, had sought to bring near the eyes of us mortals."

Galileo's will requested that he be buried next to his father, Vincenzo, in the family tomb in the Basilica of Santa Croce. However, for fear of provoking the anger of the Church, he was buried in a very small chamber under the Basilica's bell tower. The Grand Duke Ferdinando planned to construct a monumental tomb for him opposite that of the celebrated artist Michelangelo Buonarroti, but this proposal was vetoed by Pope Urban VIII, who continued to maintain that Galileo's ideas were not only false but also perilous to Christianity. In this case as well, Galileo ultimately had the upper hand. Even though his remains lay for almost a century in an obscure chamber, the last will of his most admiring disciple, Viviani, ensured that they were moved on March 12, 1737, to an impressive sarcophagus, above which an imposing monument was later erected (Figure 11 in the color insert). Viviani, in fact, devoted much of his life to the task of creating what he regarded as a fitting final resting place for his great master, and he effectively turned even the façade of his own house into a monument to Galileo (Figure 9 in the color insert). The Church's journey to recognizing its mistakes in the Galileo Affair was slower and far more tortuous.

CHAPTER 16

The Saga of Pio Paschini

Perhaps no other story than the tale of Monsignor Pio Paschini demonstrates better why Galileo's fight for freedom of thought still needs to be presented, examined, and understood today.

In 1941 the Pontifical Academy of Sciences decided to publish a new biography of Galileo for the occasion of the three hundredth anniversary of his death. The goal of the project was described by the academy's president, Agostino Gemelli, as the production of "an effective demonstration that the Church did not persecute Galileo but helped him considerably in his studies." Sensing perhaps that this statement about the expected result could elicit surprise or even shock, Gemelli added that the book "will not be a work of apologetics, because that is not the task of scholars, but will be a historical and scholarly study of the documents." Monsignor Pio Paschini, a highly respected professor of church history and rector of the Pontifical Lateran University in Rome was selected to write the book. Paschini was known for both his orthodoxy and his integrity.

Even though Paschini had no prior experience in the history of science (he admitted that the theories of the universe were "abstruse and boring" to him), nor was he in any sense a Galileo scholar, he arduously worked on the project, managing to complete the book *Life and Works of Galileo Galilei* in just three years and producing a manuscript on January 23, 1945. Protocol required that Paschini submit the book for review by the Church's authorities. That's when, ironically, history repeated itself. Since Paschini's impartial, sincere judgment of Galileo's life contained some serious criticism of the Church's behavior, the book displeased both Gemelli and the Holy Office and was rejected as "unsuitable" for publication.

An examination of Paschini's correspondence concerning the rejection, especially that with his friend Giuseppe Vale, a priest, historian, and archivist, reveals that the main reason for not approving the book was that it had been judged as being "nothing but an apology for Galileo." Paschini put the blame for Galileo's condemnation squarely at the Church's and the Jesuits' door. He explained that in the *Dialogo*, Galileo presented objectively the opinions for and against Copernicanism. It wasn't Galileo's fault, Paschini argued, that Copernicanism appeared much stronger. From Pascini's letters, we can further infer that the reviewers of his book based part of their criticism on Bellarmino's old argument—namely, that there was no conclusive proof for the Earth's motion. Paschini readily dismissed this by pointing out that there had been even less conclusive evidence for the Ptolemaic geocentric model.

Even though Paschini initially protested and fought the decision, he eventually gave up and obeyed the request not to discuss the affair any further "for the good of the Church." Paschini died in December 1962, legally leaving his unpublished manuscript to the care of his former teaching assistant Michele Maccarrone. In 1963 Maccarrone embarked on a campaign to publish the book. He conducted a series of meetings with various church officials, including Pope Paul VI, who, in his previous position as deputy secretary of state, had informed Paschini about the negative reviews of his book.

Maccarrone's efforts seemed to have borne fruit, since the Pontifical Academy of Sciences showed interest in publishing the book, this time on the four hundredth anniversary of Galileo's birth. The academy tasked Jesuit textual scholar Edmond Lamalle with updating the book. Lamalle made a series of revisions that he described as "deliberately very discreet" and "limited to corrections that seemed to us to be indispensable." He also added an introduction in which he outlined what he regarded as shortcomings of the original manuscript—weaknesses that he supposedly attempted to correct. The revised book was published on October 2, 1964, under the same title, with a preface by Lamalle. The general impression one got from Lamalle's comments was that the published book was essentially identical to Paschini's manuscript, other than some minimal editorial corrections.

Around the same time, during the Second Vatican Council, held in four annual sessions of two or three months each from 1962 through 1965, the Church was engaged in discussions concerning the relationship between religion and science, under the general theme of "the Church in the modern world." As part of that discourse, a draft of a committee report included the significant sentence: "It is necessary that we do our best, insofar as human frailty permits, that such errors [where science is presented as opposing faith], as for example the condemnation of Galileo, are never repeated." Due to opposition from a few bishops, however, this text, which mentioned the Galileo affair explicitly, was dropped in favor of a more general statement, saying:

"One can, therefore, legitimately regret attitudes to be found sometimes even among Christians, through an insufficient appreciation of the rightful autonomy of science, which have led many people to conclude from the disagreements and controversies which such attitudes have aroused, that there is opposition between faith and science."

Galileo's case was pushed to a footnote: "See P. Paschini, *Life and Works of Galileo Galilei*, 2 vols., Pontifical Academy of Sciences, Vatican City, 1964."

All of this might have still seemed moderately appropriate were it not for the fact that in 1978 Pietro Bertolla, a participant in a conference honoring Paschini, decided to compare word for word Pas-

chini's original manuscript with the published work. He discovered a
few hundred changes, which numberwise didn't seem excessive, given
the book's more than seven hundred pages. However, when Bertolla
scrutinized the individual changes, he realized that in some places
the revisions resulted in the opposite of Paschini's original meaning. In particular, Lamalle played down the significance of Galileo's
scientific discoveries and heaped more of the blame on him in his
interactions with the Inquisition. For example, when discussing the
anti-Copernican decree of 1616, Paschini wrote:

> . . . to be directed against the Copernican doctrine and to arrive
> at a condemnation in a decree pronounced with a levity that was
> wholly unusual on the part of the austere tribunals. What is worse
> is that one never revisited that decree with a weightier examina-
> tion. The Peripatetics [Aristotelian philosophers] had won and
> did not want to let go so soon of the victory. As regards Galileo,
> he was silenced by means of an injunction.

In the published book, on the other hand, Lamalle changed the text
to read: "This decree appears surprising today considering that it
came from such a balanced and austere tribunal, but it should not be
surprising if we consider it in the context of the doctrine and the scientific knowledge of that time." In other words, while Paschini maintained that the anti-Copernican decree of 1616 was at best careless,
and the fact that it had never been reexamined inexcusable, Lamalle
made it look as if Paschini said that while the decree was unfortunate, and not one that we would expect from such a wise body as the
Inquisition, it was entirely understandable for the early seventeenth
century. The important point here was not whether Lamalle was
correct in his interpretation but, rather, the intellectual dishonesty
in presenting his own views as if they were Paschini's, without mentioning this fact. A similar deception occurred in the presentation of
Paschini's conclusion about Galileo's condemnation in 1633. Citing an
article from 1906, Paschini wrote:

Regarding the responsibility [for the condemnation], one can frankly say that "the persons who are most to blame in the eyes of history are the defenders of an outdated school who saw the scepter of science slipping from their hands and could not bear that the oracles coming out of their lips should no longer be religiously listened to, and so they used all means and intrigues to regain for their teaching the credit it was losing. One of the chief means used were the Congregations and their authority, and the latter's fault was to have allowed themselves to be used."

Paschini was putting the blame on the conservative Jesuits and the Inquisition. Lamalle replaced this entire passage by citing an article from 1957 claiming that "one was dealing with a great struggle . . . scientific reason took a bold step, although without advancing decisive proofs; and such a giant step necessitated recombination of the familiar images that were connected to the representation of the universe, in the mind of the scientist as well as in that of the man on the street."

Put another way, Lamalle dismissed Paschini's view as outdated. Again, irrespective of the validity of Lamalle's perspective, in spite of his claims to the contrary, he essentially rewrote Paschini's book, at least with regard to Paschini's conclusions about the Inquisition's treatment of Galileo and Copernicanism.

The entire Paschini saga, which occurred in the middle of the twentieth century, does leave us with a bitter taste and a suspicion that the restrictions by the Church on freedom of thought, and the associated intellectual dishonesty, are still not something of the very distant past.

The story took a new turn in 1978. Pope John Paul II, who was elected that year, had previously experienced the denial of personal and religious freedom in Communist Poland, his homeland. It was therefore only to be expected that he would at some point address the interaction between science and religion in general and the Galileo affair in particular. Indeed, he did, the very next year.

THE "HARMONY BETWEEN SCIENTIFIC TRUTH AND REVEALED TRUTH"

On the occasion of the centennial of Einstein's birth, the Pontifical Academy of Sciences held a conference at which John Paul II delivered a speech entitled "Deep Harmony Which Unites the Truths of Science with the Truths of Faith." In this address, the Pontiff made a few historically important admissions. First, he acknowledged that Galileo had "had to suffer a great deal" from the actions of church officials and institutions. Second, he noted that the Second Vatican Council "deplored" unwarranted religious interventions in scientific matters. The Pope was also quick to point out that Galileo himself (in his *Letter to Benedetto Castelli* and *Letter to the Grand Duchess Christina*) expressed the view that science and religion are harmonious and not in contradiction with each other if Scripture is adequately interpreted.

Perhaps most important, the Pope encouraged a new study to be conducted of the entire Galileo affair, one that would be considered "with complete objectivity." When this initiative was announced in October 1980, it made headlines all around the world. The *Washington Post*, for instance, declared, "World Takes Turn in Favor of Galileo." The *Post*'s article itself concluded that the Pope's action was designed "to wipe out a judgment that in living memory was being used by adversaries of the Church as a symbol of its opposition to intellectual freedom." The actual charge of the appointed commission did not speak of "retrying" Galileo, but, rather, expressed the intention "to rethink the whole Galileo question."

The Vatican commission issued its final report on October 31, 1992, and the Pope acknowledged that he regarded the commission's work as done. After hearing a presentation from the commission's chairman, the Pope himself gave a speech, under the auspices of a meeting on the phenomenon of *complexity* in mathematics and science. One of his main points addressed the relationship between the results of scientific research and interpretations of Scripture—the topic to which

Galileo had devoted so much of his reasoning powers, only to be frustrated by the Church. The Pope admitted: "Paradoxically, Galileo, a sincere believer, proved himself more perspicacious on this issue than his theologian adversaries. The majority of theologians did not perceive the formal distinction that exists between the Holy Scripture in itself and its interpretation, and this led them unduly transferring to the field of religious doctrine an issue which actually belongs to scientific research."

The Pope presciently added that the lessons from the Galileo affair would probably become relevant in the future, when "one day we shall find ourselves in a similar situation." He then repeated his belief that science and religion are in perfect harmony.

With that presentation, the Church basically declared the Galileo case closed. Media around the world had a feast. The *New York Times* announced: "After 350 Years, Vatican Says Galileo Was Right: It Moves." The *Los Angeles Times* had a similar message: "It's Official: The Earth Revolves Around the Sun, Even for the Vatican." Some Galileo scholars were not amused. Spanish historian Antonio Beltrán Marí wrote: "The fact that the Pope continues to consider himself an authority capable of saying something relevant about Galileo and his science shows that, on the Pope's side, nothing has changed. He is behaving in exactly the same manner as Galileo's judges, whose mistake he now recognizes."

To be fair, the Pope was in some sense in a no-win situation. Whatever he said, or failed to say regarding the Church's errors would have been criticized on some grounds. Nevertheless, the chiefly theological rehabilitation of Galileo was far too late.

Interestingly, in both his 1979 speech and the one in 1992, Pope John Paul II mentioned Albert Einstein. In 1979 he started his address with: "The Apostolic See wishes to pay to Albert Einstein the tribute due to him for the eminent contribution he made to the progress of science, that is, to knowledge of the truth present in the mystery of the universe." This led the Pope to the following conclusion: "Like any other truth, scientific truth is, in fact, answerable only to itself and to the supreme Truth, God, the creator of man and of all things." In his 1992 presentation, he reiterated the same idea. Starting with a popular form in which an aphorism by Einstein is often quoted—"What is eternally incomprehensible in the world is that it is comprehensible"—the Pope suggested that the intelligibility "leads us, in the final analysis, to that transcendent and primordial thought imprinted on all things."

Given the frequent references to Einstein, it is interesting in this discussion of the relationship between science and religion to examine Einstein's thoughts on religion and God, and to compare those with Galileo's, more than three centuries earlier.

CHAPTER 17

Galileo's and Einstein's Thoughts on Science and Religion

In his *Letter to the Grand Duchess Christina*, Galileo expressed in the clearest fashion what he regarded as the proper relationship between science and religion. This document was, at the same time, a manifesto of Galileo's fight for intellectual freedom—for the right of scientists to defend what they view as compelling evidence. One of the reasons for Galileo's clash with the Church had to do with the rather different interpretations he and church officials had given to the actual nature of the disagreement. Whereas Galileo was convinced that he was trying, in some sense, to save the Church from committing a monumental error, these officials treated his obstinate insistence on the validity of his opinions as a direct attack on the sanctity of Scripture and on the Church itself. To support his point of view, Galileo recruited the writings of Saint Augustine, who cautioned against making conclusive statements about things that are hard to understand: "We ought not to believe anything inadvisedly on a dubious point, lest in favor to our error we conceive a prejudice against something that truth hereafter may reveal to be not contrary in any way to the sacred books of either the Old or the New Testament." Basically, Saint Augustine was arguing in the fifth century that biblical texts should not be understood literally, if they contradict what we know from reliable sources. This was precisely the mistake, Galileo contended, that his adversaries were committing: "They would extend such authorities [of the Bible and Holy Fathers] until even in purely physical matters—where faith is not involved—they would have us altogether abandon reason and the evidence of our senses in favor of some biblical passage, though under the surface meaning of its words, this passage may contain a different sense."

Galileo reiterated this idea repeatedly, expressing his conviction that "He [God] would not require us to deny sense and reason in physical matters which are set before our eyes and minds by direct experience or necessary demonstrations." Referring more specifically to his Copernican convictions, Galileo was emphatic that "the mobility or stability of the Earth or Sun is neither a matter of faith nor one contrary to ethics." You might have thought that more than three centuries later, we would not have to go through some of the same types of adversities that Galileo experienced when it came to literal interpretations of the Bible, but unfortunately that is not the case. For example, a 2017 Gallup survey in the United States found that about 38 percent of adults were inclined to believe that "God created humans in their present form, at one time within the last ten thousand years."

DARWIN VERSUS "INTELLIGENT DESIGN"

All of Galileo's arguments are equally valid for the topic of teaching Darwin's theory of evolution, the controversy over which remains as lively today as it has ever been. Astonishingly, even though the Pope himself admitted that it had been wrong to transfer "to the field of religious doctrine an issue which actually belongs to scientific research," and in spite of more than a century of solid evidence confirming evolution by means of natural selection, many Americans—and a considerable number of people in other parts of the world—still adhere to creationist ideas. Sadder yet, the beliefs on the creationist side have been so strong that there appears to be no resolution in sight in popular opinion and the dispute over how the subject should be taught at school continues. *One cannot emphasize enough the fact that there isn't even the slightest scientific doubt.*

First, the age of the universe is now known with an uncertainty of less than ten percent. Second, the National Academy of Sciences clearly pronounced: "The concept of biological evolution is one of the most important ideas ever generated by the application of scientific methods to the natural world." On October 27, 2014, Pope Francis issued a statement at the Pontifical Academy of Sciences that said: "The big bang, which nowadays is posited as the origin of the world, does not contradict the divine act of creating, but rather requires it. The evolution in nature is not inconsistent with the notion of creation, as evolution presupposes the creation of beings that evolve." He was following here in the footsteps of Pope John Paul II. The latter said about the theory of evolution in an address on October 22, 1996: "It is remarkable that this theory has had progressively greater influence on the spirit of researchers, following a series of discoveries in different scholarly disciplines. The convergence in the results of these independent studies—which was neither planned nor sought—constitutes in itself a significant argument in favor of the theory."

In spite of these clear judgments from the highest scientific and religious authorities, quite a few people simply refuse to be convinced. Even more embarrassingly, creationists have occasionally managed to convince educators, politicians, and judges that evolution is merely a "theory," and that the more fashionable variant of creationism— "intelligent design"—should be taught side by side with evolution, in science classes.

The similarity between the arguments by creationists and those advanced against Galileo is absolutely striking. First, creationists say, evolution by means of natural selection is not a proven fact, and it hinges on processes that were not observed, nor can they ever be observed. To this, biologists respond that the fossil record does provide ample compelling evidence that organisms have evolved over the age of the Earth. In fact, the theory of evolution could have been falsified easily had it been incorrect (which is one of the hallmarks of an acceptable scientific theory). For instance, finding even one fossil of an advanced mammal, such as a mouse, dating to two billion years ago would have been sufficient to refute the entire theory. No such evidence has ever been found. On the contrary, the findings fully support evolution. For example, evolution predicts that from the period between a few million years ago and a few hundred thousand years ago, we should find fossils of hominins (ancestors of modern humans) with progressively less apelike features. This prediction has been unambiguously confirmed. Furthermore, no fossils of anatomically modern humans dating to millions of years ago have ever been discovered. We should also note that there are numerous examples of natural selection at play, ranging from bacteria becoming resistant to certain types of antibiotics, to the evolution of the colors of the peppered moth in nineteenth-century England.

A second objection that creationists raise is the claim that no transitional fossils, such as half-bird-half-reptile creatures have been found. This is simply false. Paleontologists have discovered fossils that are intermediate between taxonomic groups. For example, a fossil named *Tiktaalik roseae*, which dates to about 375 million years ago, demonstrates the transition from fish to the first legged land animals, and a whole series of fossils chronicles the transition from a small species called *Eohippus* to today's horse, over a geologic timescale of about fifty million years.

Finally, creationists resort to an argument the origins of which date all the way back to the Roman orator Cicero in the first century BCE: such complex "machines" as exhibited by various life forms, the claim goes, could only have been produced by an "intelligent design." In the early nineteenth century, natural theologian William Paley adopted the same line of reasoning: an intricate watch attests to the existence of a watchmaker. Creationists latched particularly on the example of the eye as an anatomical organ that could not have evolved naturally. However, the discovery of more primitive organs that trace the evolution of light-sensing apparatus has invalidated this argument as well. Fundamentally, any biological feature that appears to be a contrivance was the outcome of a long evolutionary selection aided by symbiosis with the environment. In general, processes that are not fully understood don't constitute flaws. Creationists seem to forget or ignore that Galileo had already fought this type of battle four centuries ago, and eventually won.

The continuing debate concerning climate change is even worse, since to avoid catastrophic consequences requires a much more rapid response. Climate change denial is fed mainly by political, financial, and religious motivations. Unlike Darwinian evolution, where rejection of the theory is strongly correlated with religiosity, in the case of climate change, political conservatism is the dominant cause for denial. The religious component was perfectly captured by what Senator James Inhofe of Oklahoma told the Voice of Christian Youth America's radio program *Crosstalk with Vic Eliason* in 2012: "God's still up there. The arrogance of people to think that we, human beings, would be able to change what He is doing in the climate is to me outrageous." Contrast this with the fact that there is now overwhelming expert scientific consensus (about 97 percent) that "it is extremely likely that human influence has been the dominant cause of the observed warming since the mid-twentieth century."

The United Nations Environment Program's Emissions Gap Report in 2018 showed that carbon dioxide (CO_2) emissions actually rose in 2017—the first time after stalling for four years. This is particularly alarming in view of the latest report by the Intergovernmental Panel on Climate Change (IPCC)—a group of scientists convened by the UN to guide world leaders—which concluded that to limit the global temperature rise to 2.7°F (1.5°C) above preindustrial levels would require slashing emissions of greenhouse gases by 45 percent by 2030. In an unprecedented move, on December 2, 2018, the presidents of four former UN climate talks issued a joint statement calling for urgent action. The fact that the United States has withdrawn from the Paris Agreement on Climate Change, and President Donald Trump's continued promotion of fossil fuels are, in this context, nothing short of shocking (as is Trump's response to the COVID-19 pandemic). As Nobel laureate in physics Steven Weinberg put it: "It is generally foolish to bet against the judgments of science, and in this case, when the planet is at stake, it is insane."

Any list of the top scientists in history—those who changed the world—includes the names of Galileo and Einstein. This is an additional reason why, when discussing the topic of science and religion, it is interesting to compare the views of these two geniuses. We know that Galileo regarded Scripture as the guide to faith, ethics, and moral behavior (for "salvation"), and objected to literal interpretations of biblical texts only when they contradicted scientific observations and logical demonstration. More than three centuries later, Einstein shared Galileo's scientific perspective, but he was almost diametrically opposed to him on matters of faith.

EINSTEIN ON RELIGION AND SCIENCE

There is no doubt that on the topic of intellectual freedom, the German-born Einstein had precisely the same views as Galileo. In a 1954 statement he made for a conference of the US Emergency Civil Liberties Committee, Einstein said: "By academic freedom, I understand the right to search for truth and to publish and teach what one holds to be true." He was repeating here his own thoughts from an address written in 1936, three years after Adolf Hitler came to power in Germany—and Einstein immigrated to the United States: "Freedom of teaching and of opinion in book or press is the foundation for the sound and natural development of any people." Galileo certainly would have agreed.

On the relation between science and religion, on the other hand, Einstein's opinions were more complex. Here is a very brief synopsis.

Einstein mentioned God quite frequently in his writings, in talks, and in conversations. For example, when he wanted to pronounce his skepticism about quantum mechanics, the theory of the subatomic world, he said famously, "He [God] does not play dice." Similarly, when Einstein expressed the opinion that nature may be difficult to decipher, but it is not bent on trickery, he said: "The Lord God is subtle, but malicious he is not." Einstein even wondered whether the cosmic blueprint had allowed for any choice: "What really interests me is whether God could have created the world any differently; in other words, whether the requirement of logical simplicity admits a margin of freedom." These quotes, however, are related mostly to the structure of the universe, and they don't give us the full picture of Einstein's attitude toward religion.

Einstein developed most of his perspective on religion, science, and the interaction between the two in a series of essays, letters, and speeches written mainly between 1929 and 1940. One of the first of these, an essay entitled "What I Believe" written in 1930, contains some of Einstein's most memorable quotes:

> The most beautiful experience we can have is the mysterious. It is the fundamental emotion which stands at the cradle of true art and true science. Whoever does not know it and can no longer wonder, no longer marvel, is as good as dead, and his eyes are dimmed. It was the experience of mystery—even if mixed with fear—that engendered religion. A knowledge of the existence of something we cannot penetrate . . . it is this knowledge and this emotion that constitute true religiosity. In this sense, and only this sense, I am a deeply religious man. I cannot conceive of a God who rewards and punishes his creatures, or has a will of the kind that we experience in ourselves.

Einstein repeats here an opinion he expressed in 1929, in response to a telegram from Rabbi Herbert Goldstein, in which the rabbi asked him, "Do you believe in God?" As a lifelong admirer of the seventeenth-century Dutch Jewish rationalist philosopher Baruch Spinoza, Einstein replied: "I believe in Spinoza's God, who reveals himself in the lawful harmony of the world, not in a god who concerns himself with the fate and the doings of humankind."

Einstein further expanded on these views in two articles, one entitled "Religion and Science," which he wrote for the *New York Times Magazine* in November 1930, and the other entitled "Science and Religion," which was read at a conference in New York in 1940. In the former, Einstein outlined what he regarded as the three main steps in the evolution of religious beliefs, while in the latter he attempted to define science and religion, and expressed his view as to what he thought was the basic source of the perceived conflict between the two.

The three phases in the development of religions, according to Einstein, were: fear ("of hunger, wild beasts, sickness, and death"); the "social or moral conception of God" (a God that rewards, punishes, and comforts); and the "cosmic religious feeling." Einstein himself admitted to feeling only this third religious experience:

"The man who is thoroughly convinced of the universal operation of the law of causation cannot for a moment entertain the idea of a being who interferes in the course of events—provided, of course, that he takes the hypothesis of causality really seriously. He has no use for the religion of fear and equally little for social or moral religion."

Evidently, for Einstein religion played a very different role from the one it played for Galileo. Whereas both agreed that nature operates according to certain mathematical laws, as we have seen, Galileo regarded Scripture as the chief guide for moral behavior that eventually leads to salvation, while Einstein's religious sensation was inspired precisely and solely by those laws of nature.

Einstein's definitions of science and religion endeavored to go even further. He defined science as "the attempt at the posterior reconstruction of existence by the process of conceptualization." That is, science, according to Einstein, describes reality as it exists and not what reality should ideally be. Religion, on the other hand, Einstein explained, was "the age-old endeavor of humankind to become clearly and completely conscious of these values and goals [to be liberated from selfish, ego-oriented desires and to have superpersonal aspiration for the improvement of existence] and constantly to strengthen and extend their effect." Specifically, to Einstein, traditional religion ascertains a *desired* state rather than reality. From these two definitions Einstein concluded that there shouldn't have been any clash between science and religion *unless* religious institutions intervened into the realm of science (for instance, through insisting on literal interpretations of the Bible, as in the cases of Galileo and Darwin), or through the introduction of the doctrine of a "personal God," which to Einstein was scientifically unacceptable.

While Einstein admitted that science does not possess the tools to unequivocally refute the concept of a personal God, he regarded this notion as "unworthy," because it could "maintain itself not in clear light but only in the dark."

It was Einstein's denial of a personal God that provoked an extremely strong reaction, mostly negative, from many circles. In a language closely resembling that of the Aristotelians against Galileo, a priest from North Hudson, New York, wrote in the *Hudson (N.Y.) Dispatch*: "Einstein does not know what he is talking about. He is all wrong. Some men think that because they have achieved a high degree of learning in some field, they are qualified to express opinion in all."

Monsignor Fulton John Sheen, a professor at the Catholic University of America, criticized both the essay from 1930 and the one from 1940, concluding sarcastically, "There is only one fault with his [Einstein's] cosmical religion, he put an extra letter in the word—the letter 's'."

Not all the reactions were negative. A disabled World War I veteran from Rochester, New York, wrote: "The great leaders, thinkers, and patriots of the past who fought and died for free thought, free speech, free press, and intellectual liberty arise to salute you! With the great and mighty Spinoza, your name will live as long as humanity."

Einstein himself was particularly annoyed by the fact that he had been labeled an atheist. At a charity dinner in New York, he told an anti-Nazi German diplomat in words very reminiscent of Viviani's eulogy of Galileo: "In view of such harmony in the cosmos which I, with my limited human mind, am able to recognize, there are yet people who say there is no God. But what really makes me angry is that they quote me for support of such views."

Fascinatingly, a letter that Einstein wrote on January 3, 1954, in which he repeated his views about a "personal God" and "Spinoza's God" sold at an auction at Christie's for the astounding sum of $2,892,500 on December 4, 2018. The letter was addressed to the German Jewish philosopher Eric Gutkind, in response to a book Gutkind had written, which was a religious, humanistic manifesto based on biblical teachings. Perhaps the most significant sentiment expressed in the letter was Einstein's agreement with Gutkind that humans should aspire for "an ideal that goes beyond self-interest, with the striving for release from ego-oriented desires, the striving for the improvement and refinement of existence, with an emphasis on the purely human element."

Given that the relation between science and religion is a topic that will likely continue to be discussed by future generations, one piece of advice from Galileo and Einstein appears to be particularly insightful and helpful. *As long as the conclusions of science concerning physical reality are accepted, with no intervention of religious beliefs and no denouncing of provable facts, no conflict between the two realms can exist.* Galileo understood that the Bible is not a science book. It represents an allegorical description of the awe that humans in antiquity felt when faced with a seemingly incomprehensible universe. Einstein still felt the same awe, even though he concluded that the cosmos was comprehensible after all. This was, in some sense, also the judgment of Pope John Paul II. A peaceful coexistence between science and mainstream religion (I exclude here both religious fanatics and aggressive "missionary" atheists) is therefore definitely possible, at least in principle. Philosopher of science Karl Popper nicely expressed his moderate views on this topic by writing: "Although I am not for religion, I do think that we should show respect for anybody who believes honestly." Nevertheless, recognizing that the risk of conflict remains, the Pope's suggestion to engage in "a dialogue in which the integrity of both religion and science is supported and the advance of each is fostered" appears to be a good step forward. We should allow for the coexistence of many ideas and ideals and the freedom to debate those, and disallow only intolerance.

CHAPTER 18

One Culture

Galileo would not have understood C. P. Snow's concept of "two cultures." The idea that literary or humanistic intellectuals would form a distinct group that excludes scientists and mathematicians would have been foreign to him. He himself comfortably inhabited both worlds, the scientific and the humanistic, commenting on art and literary works and performing music with the same passion and enthusiasm that characterized his scientific work. His artistic training both informed his interpretation of observations and allowed him to present his discoveries more effectively. Moreover, in publishing much of his work in witty but easy to understand Italian rather than Latin, Galileo was also a perfect example of what author John Brockman dubbed a "third-culture thinker"—a person who disposes of any mediators and communicates directly with the intelligent public.

If you think about it for a moment, how is it even possible not to consider science as an integral part of the human culture and intellectual heritage? Science is, after all, a realm in which one can point to unambiguous progress. It would be difficult to argue convincingly that, say, art today is distinctly superior to Renaissance art, or that Sappho's poetry was clearly inferior to that of Emily Dickinson. On the other hand, life expectancy in England in the seventeenth century was about thirty-five years, while now, mainly as a result of undertakings emerging from science, it is (averaging over women and men) about eighty-one. Or take the fact that Galileo was the first person to characterize correctly features on the lunar surface, while by now a dozen astronauts have walked on the surface of the Moon. Similarly, Antonie van Leeuwenhoek, whose life partly overlapped with Galileo's, established microbiology as a scientific discipline and identified microbes as new species. Since then, however, millions of biological species have been fully characterized. Finally, whereas Galileo was accused that his investigations into the nature of matter conflicted with biblical descriptions, today's particle physicists have managed to discover all the basic constituents of ordinary matter. And this list of scientific achievements just goes on and on, with incredible advances made both in the exploration of the physical and biological microcosms and the cosmic macrocosm. Isn't this progress an essential part of just one human culture?

Imagine that we had to communicate now with an alien superior Galactic civilization. What would be a succinct way to convey to them (assuming that they could understand us) the intellectual and technological levels of our civilization? Believe it or not, one interesting and relatively straightforward method would be to simply inform them that we have succeeded in detecting gravitational waves from the collision of two black holes. Why would this seemingly esoteric topic provide such a powerful and informative statement? Gravitational waves are ripples in the fabric of space-time produced by large accelerations, such as those created in the case of two neutron stars or two black holes that are spiraling toward each other. These waves were predicted to exist in Einstein's theory of general relativity—a theory that changed the description of gravity from a mysterious force that acts across distance to one representing the curvature of space-time. That is, just as a heavy object causes a trampoline to sag, masses (such as the Sun or a black hole) warp space-time in their vicinity. When those masses accelerate, the disturbance propagates as ripples. On the theoretical side, therefore, informing an alien civilization that we know about gravitational waves communicates immediately the status of our understanding of the nature of space-time, a crucial element in the evolution of our universe. The fact that we have managed even to *detect* these gravitational waves conveys an immediate status report on our technological prowess, since the ability to detect these extraordinarily weak ripples is nothing short of miraculous. Basically, the gravitational wave researchers sensed a wave that stretched space-time by 1 part in 1,000,000,000,000,000,000,000. That is, this wave made the entire Earth expand and contract by about the width of an atomic nucleus.

There surely aren't many people who don't acknowledge that scientific progress has been responsible for many of the improvements in the quality of our everyday lives. Unfortunately, the achievements in the humanities don't always receive the respect they deserve—a phenomenon that would have undoubtedly upset Galileo.

The contribution of the humanities to our ability to imagine even things that don't exist, to human creativity, and to the development and evolution of the human language, with all of its associated ramifications in communication, cannot be overemphasized. Philosophy, soul-searching, and religion have helped humans in the construction of a moral framework. Still, there are those who instead of fostering a partnership between the sciences and the humanities—one which takes advantage of the assets that each domain has to offer—insist instead that the sciences and the humanities should strive for a hermetically separate coexistence. These "separatists" call for clear (if somewhat porous) borders. In my humble opinion, while certainly there are important differences in the subjects, styles, and practices of the two domains, the recognition that both the humanities and the sciences are integral parts of one human culture must come from both sides.

This conclusion becomes particularly obvious when we realize that a few of the most fundamental questions that humans have ever posed, have crossed, over the millennia, first the boundary between religion and philosophy, and later, the border between philosophy and science. I am referring in particular to questions related to *origins*: How did the universe begin? How was the Earth formed? How did life on Earth start? How did consciousness emerge? These, as well as perhaps the even bigger question of Why does the universe exist?— or, as it is sometimes phrased, Why is there something rather than nothing?—are now widely recognized (although unfortunately not by all) as belonging to the science domain. More important, science has already provided at least partial answers to some of these questions.

For example, we now know that our universe started about 13.8 billion years ago, from an extremely hot and dense state, commonly known as the big bang. We can almost determine the age of the universe with a higher precision than we can determine the age of a living person. We know that the Sun formed 4.6 billion years ago from the gravitational collapse of a cloud of gas and dust, and that the Earth formed through the coalescence of dust particles in the flat disk that formed around the Sun, and so on. Just as Galileo predicted, much of our picture of the world is based not on vague, qualitative descriptions but on detailed mathematical models and numerical simulations.

These facts only increase the urgency of fighting illiteracy on *all* fronts, the scientific and the humanistic. In the same way that everyone should at least have the opportunity to be exposed, for instance, to a few of Shakespeare's plays, or the writings of Marcel Proust, F. Scott Fitzgerald, Virginia Woolf, Chimamanda Ngozi Adichie, Lu Min, and Fyodor Dostoevsky, he or she should also be aware that the world is governed by certain laws of nature, and that there is compelling evidence that these laws apply to the entire observable universe, and they don't appear to change with time.

As you may recall, Galileo objected vehemently to any type of compartmentalization, be it of different branches of science or even between the sciences and mathematics or the arts. He regarded such hermetic separations as "no less foolish than that of a certain physician who, moved by a fit of spleen, said that the great doctor Acquapendente [the sixteenth-century Italian surgeon Girolamo Fabrizio], being a famous anatomist and surgeon, should content himself to remain among his scalpels and ointments without trying to effect cures by medicine." Galileo would have undoubtedly resisted any attempt to exclude either the humanities or the sciences from being an indispensable part of human culture. The point is that human culture is diverse. The essence of this fact has been captured in one sentence by University of Chicago philosopher Martha Nussbaum, when she said: "Education needs to impart skills of critical thinking, and it needs to cultivate the imagination." These are indeed the crucial elements provided by the sciences and the humanities. Science attempts to *explain* and predict the universe. Literature and the arts provide *our emotional response* to it. Concepts such as freedom of thought emerge from the fusion of these disciplines. Galileo understood centuries ago that humans need both the humanities and the sciences. It is fitting that Galileo—one of the greatest scientists in history—has been immortalized in so many works of art (Figure 12 in the color insert shows a bust of Galileo by Carlo Marcellini). That is why, perhaps, the last words Bertolt Brecht put in the mouth of the blind astronomer in the play *Life of Galileo* were the poignant, "What's the night like?"

Acknowledgments

I am deeply indebted to many people and institutions who helped me bring this project to fruition. I would like to thank the Museo Galileo in Florence, Italy, and its staff for their kind hospitality. I am grateful to the museum's director, Paolo Galluzzi, and vice director, Filippo Camerota, for very helpful discussions on Galileo, and to Giorgio Strano for conversations on Tycho Brahe. Thanks to Giulia Fiorenzoli for her help in facilitating my stay. Alessandra Lenzi, Elisa Di Renzo, Sabina Bernacchini, and Susanna Cimmino greatly assisted me in the museum's library and provided materials from the photo lab. I am especially grateful to Galileo scholars Michele Camerota and Maurice Finocchiaro for fascinating conversations on Galileo and for providing me with some of their important publications. I thank Federico Tognoni for his assistance with Galileo iconography. I had stimulating discussions with philosopher of science Dario Antiseri on philosophy, and on the relationship between science and religion, and science and the humanities. Stefano Gatti (who sadly passed during the writing of this book) and Marquis Mariano Cittadini Cesi provided me with important information on Galileo's friend and patron Federico Cesi.

Geologist and environmental scientist Daniel Schrag explained to me in detail the science of climate change and introduced me to important articles on the topic. Atmospheric physicist Richard Lindzen, who is one of the more vocal climate change "deniers," explained to me what precisely he objects to in the interpretation of climate change findings.

I am grateful to Amy Kimball at the Special Collections of the Sheridan Libraries at Johns Hopkins University for providing me with crucial materials. Kate Hutchins at the Special Collections Research Center at the University of Michigan provided me with an invaluable original document by Galileo.

Art historian Lisa Bourla gave me important information on painter and Galileo's friend Cigoli. Curators Joost Vander Auwera and Ingrid Goddeeris from the Royal Museum of Fine Arts in Brussels helped me to search for a portrait of Galileo entitled *Galileo in Prison.* Art experts Benito Navarrete Prieto, Pablo Hereza, Jonathan Brown, and Xanthe Brooke gave authoritative opinion on the attribution of that Galileo portrait. Curators Annemie De Vos from the Vleeshuis Museum in Antwerp and Els Baetens from the Stedelijk Museum in Sint-Niklaas gave invaluable information and greatly assisted in the research concerning the whereabouts of the same portrait.

My wife, Sofie Livio, showed an infinite patience and continuous support during the years of research for and writing of this book. For these, I am forever grateful.

Finally, my agent, Susan Rabiner, encouraged me to write the book and skillfully guided me throughout the writing process. I thank Sharon Toolan for her professional assistance in preparing the manuscript for print. I am deeply indebted to my editor, Bob Bender, for his thoughtful comments on the manuscript, and to Johanna Li and the entire production team at Simon & Schuster for their support in the preparation of the book for publication.

Notes

The main source of information on Galileo's life and work has been (since the beginning of the twentieth century) Antonio Favaro's monumental *Le Opere di Galileo Galilei: Edizione Nazionale* (Florence, Italy: Giunti-Barbera, 1890–1909). It was reprinted in 1929. The first edition is now available online at www.galleco.fr, and the site at *Liber Liber* (www.liberliber.it) includes much of the text. The "Galileo Project" by Albert Van Helden and Elizabeth Burr at Rice University (galileo.rice.edu) provides excellent hypertextual information. Stefano Gattei's very recent *On the Life of Galileo* presents an invaluable collection of early biographies and other important documents.

When I cited text in English, I used mostly the translations by Stillman Drake, Maurice Finocchiaro, Albert Van Helden, John L. Heilbron, Mario Biagioli, Giorgio de Santillana, Mary Allen-Olney, Stefano Gattei, Richard Blackwell, William Shea, and David Wootton.

CHAPTER 1: REBEL WITH A CAUSE

1. *At a breakfast that took place*: The event is described in more detail in chap. 5.
1. *from the book of Joshua*: Joshua 10:12–13, *NIV: Study Bible* (Grand Rapids, MI: Zonderini).
1. *letter to Castelli*: A translation of the slightly revised letter appears in Finocchiaro 1989, 49–54. See chap. 6.
3. *"the greatest of the founders of modern science"*: Russell 2007, 531.
3. *Max Born . . . once put it*: Born 1956.
5. *what we call today the scientific method*: A good discussion of Galileo's contribution is in Gower 1997, 21.
5. *Art historian Heinrich Wolfflin*: Wolfflin 1950, cited also in Machamer 1998.

5. *Galileo scholar Giorgio de Santillana*: Of particular interest is Santillana 1955, which attempts to follow Galileo's mental journey.
6. *"Those who study the ancients"*: Leonardo da Vinci, cited in Nuland 2000.
6. *Viviani further tells us*: Viviani 1717. An English translation that includes other early biographies, documents, and annoations is Gattei 2019.
6. *Giorgio Vasari, in his biographies*: Vasari 1550.
6. *Galileo himself played the lute*: On Galileo's interest in music, see Fabris 2011.
6. *Galileo could recite at length*: Good descriptions of Galileo's love for literature and art are in Panofsky 1954 and Peterson 2011.
7. *All of these discoveries created*: Machamer 1998 gives a brief, good summary of the background against which Galileo was operating. A superb description of the entire scientific culture of the time is in Camerota 2004.
8. *Was the rise of individualism*: Russell 2007 explains this trend well.
9. *The invention of movable type*: Described nicely in Eisenstein 1983. The transport of information is discussed in Reeves 2014.
11. *"the discovery of the world and of man"*: In Michelet 1855, vol. 7–8 *Renaissance et Réforme*.
12. *Einstein said about this book*: Einstein 1953.
13. *"A good many times I have been present"*: Snow 1959. The talk was delivered on May 7, 1959, in the Senate House in Cambridge. In 1963 Snow published an expanded version entitled *The Two Cultures: A Second Look*, in which he was more optimistic about bridging the gap between the two cultures.
14. *"History of Science, far from serving"*: Wootton 2015, 16.
14. *John Brockman introduced the concept*: Brockman 1995. Originally published online in 1991 at *Edge* (edge.org).
15. *"the unreasonable effectiveness of mathematics"*: Wigner 1960.

CHAPTER 2: A HUMANIST SCIENTIST

17. *was born in Pisa on February 15 or 16*: Galileo's date of birth is given more often as February 15, 1564, but two horoscopes for himself in his own hand are for February 16 and only one for February 15. Swerdlow 2004 gives a nice discussion of the horoscopes.
17. *Vincenzo apparently also became a part-time*: This is not certain. Vincenzo did accept part of Julia's dowry in the form of clothes.
17. *had two more sons and three*: Galileo's brothers were Benedetto and Michelangelo. His known sisters were Virginia, Anna, and Livia. It is unclear whether Lena was also a sister or a domestic. In *Opere di Galileo Galilei*, Vol. 19, Documenti.
18. Dialogue on Ancient and Modern Music: The book by Vincenzo Galilei was published in Florence in late 1581 or early 1582. A translation is V. Galilei 2003.
19. *the idea of using pendulums for measuring time*: This interesting speculation was raised by Drake 1978.
19. *second biographer, Niccolo Gherardini*: He was Galileo's neighbor in Rome in 1633. In conversations with Galileo, Gherardini gathered some biographical material, which he later summarized.

20. *in September 1580*: Almost all biographies give 1581, but Camerota and Helbing 2000 demonstrate convincingly that it was 1580.

20. *Tuscan court mathematician Ostilio Ricci*: Ricci became mathematician to Grand Duke Ferdinando I, but this was at a later period.

21. *"If Euclid failed to kindle"*: Einstein 1954.

21. *"One could travel securely"*: Cited in Peterson 2011.

22. *"How is it possible that mathematics"*: Einstein 1934.

23. *a small scientific tract entitled* La Bilancetta: An English translation of the essay is in Fermi and Bernardini 1961.

24. *Treatise on the Sphere, or Cosmography*: This treatise was published posthumously by the priest Urbano d'Aviso.

25. *mapping Dante's disorienting description*: The mathematical "plans" for hell are beautifully discussed in some detail in Heilbron 2010.

28. *Galileo wrote in* Two New Sciences: Galilei 1638.

29. *"If being a prodigy is not a requirement"*: Csikszentmihalyi 1996.

CHAPTER 3: A LEANING TOWER AND INCLINED PLANES

31. *as a professor and chair of mathematics*: A good description of Galileo's Pisan studies is in Wallace 1998.

32. *later biographers and historians just kept adding*: English language and literature professor Lane Cooper collected some of these stories and discussed the Leaning Tower experiment. His work has been criticized in the past, but it remains an honest effort to examine experiments in free fall. Cooper 1935. Michael Segre expertly reviews the story in Segre 1989. Camerota and Helbing 2000 beautifully discuss the background.

32. *Most present-day historians of science think*: Renowned Galileo scholar Stillman Drake thought that the demonstration did take place. Drake 1978.

33. *at the forefront of both experimental and theoretical investigations*: In a series of highly influential works, science historian Alexandre Koyré argued that Galileo could not have obtained the experimental results he described later in *Two New Sciences* using his equipment (for example, Koyré 1953, 1978). These claims have been completely refuted by Thomas Settle (Settle 1961), James MacLachlan (MacLachlan 1973), and Stillman Drake (Drake 1973). See also Clavelin 1974. The distinction between Galileo's *experientia* (experience in general) and *periculum* (experiment or test) is discussed in Schmitt 1969.

33. *This peculiar result has been shown*: For example, Thomas Settle repeated the experiment in front of a camera. Settle 1983.

34. De Motu *marked the beginning*: Galileo 1590; trans. Drabkin and Drake 1960 and Camerota and Helbing 2000 give excellent descriptions of Galileo's and other Pisan professors' ideas and experiments on falling bodies. See also Wisan 1974.

35. *'A hundred-pound iron ball'*: Galilei 1638.

35. *Apollo 15 astronaut David Scott*: Can be seen at NASA online, last modified February 11, 2016, https://nssdc.gsfc.nasa.gov/planetary/lunar/apollo_15_feather_drop.html.

35. *"When a person has discovered the truth"*: Cited in Drake 1978.

36. *his interest in philosophy*: Wallace 1998, Lennox 1986, and McTighe 1967.

37. *Here are a few lines from the controversial poem*: The poem was translated into English by astronomer Giovanni Bignami. Bignami 2000.

39. *from 1603 to 1609*: Geymonat 1965 and Heilbron 2010 give excellent descriptions of the Paduan period in Galileo's life.

39. *a few of his groundbreaking results*: Weinberg 2014 gives an excellent description of the importance of Galileo's pioneering experiments.

39. *"In a small height"*: In *Opere di Galileo Galilei*, Vol. 8, p. 128, cited in Drake 1978, 85.

40. *"Clearly a statement cannot be tested"*: Eddington 1939.

41. *The research concerning climate change*: Michael Mann has given an excellent description of the problems involved in a series of publications. Mann 2012a is a must-read. Another clear exposition is Romm 2016.

41. *Casting occasional horoscopes for students*: Galileo's totally dismissive opinion on astrology is mentioned in a letter from the archbishop of Siena, Ascanio Piccolomini, to his brother Ottavio on September 22, 1633. Bucciantini and Camerota 2005.

42. *One in particular, Gianfrancesco Sagredo*: Nick Wilding wrote a brilliant book about Sagredo. Wilding 2014.

42. *If I sometimes speculate about science*: In *Opere di Galileo Galilei*, vol. 12, pp. 43–44.

42. *in addition to being a prelate, a historian*: David Wootton wrote a captivating account about the fascinating personality of Paolo Sarpi. Wootton 1983.

42. *of the processes involved in vision*: An interesting book on the role of vision in Galileo's discoveries is Piccolino and Wade 2014.

43. *the geometric and military compass*: Galileo's instrument allowed one to perform arithmetic calculations and geometrical operations. The story of the compass is described, for example, in Bedini 1967. The website of the Galileo Museum in Florence contains beautiful images of the instrument and the booklet about it. Galileo instructed dignitaries such as Prince John Frederick of Alsace and the Archduke Ferdinand of Austria how to use the instrument. See "Instruments: The Tools of Science," www.museogalileo.it/it/biblioteca-e-istituto-di-ricerca/pubblicazioni-e-convegni/strumenti.html.

46. *Two elements in particular stand out*: Maurice Clavelin, in *The Natural Philosophy of Galileo*, referred to Galileo's contribution as the "geometrization of motion." In other words, not only that motion is interpreted in terms of quantitative laws but also that an entire body of established theorems and propositions is presented as a coherent whole. Clavelin 1974.

47. *The first letter, dated May 30, 1597*: Galileo added that if Mazzoni was satisfied with Galileo's argument, "the opinion of those great men [Pythagoras and Copernicus] and my own belief may not remain desolated."

49. *after having read only the preface of the book*: Because Kepler's emissary, Paul Hamberger, was about to immediately return to Germany. Rosen 1966.

51. *Galileo bothered to answer only a few*: One of those, Antonio Lorenzini, was expressing the doubts of philosopher Cesare Cremonini about the validity of parallax determinations. However, Lorenzini's criticism was on technical grounds—an area in which his knowledge was minimal. Galileo replied only because Kepler urged Italian astronomers to do so.

51. *composed with friends and published*: The pseudonym was Cecco di Ronchitti, and the dialogue was written in a dialect common in the Paduan countryside. The entire episode is described in detail by Heilbron 2010, 123–25.

51. *"To obtain any salary from a Republic"*: *Opere di Galileo Galilei*, vol. 10, p. 233.

52. *"I desire that the primary intention"*: In a letter to Belisario Vinta, the Tuscan state secretary to the grand duke.

CHAPTER 4: A COPERNICAN

53. *"About 10 months ago"*: Excellent translations of the *Sidereal Messenger* are by Drake 1957, 27, and Van Helden 1989.

53. *"that a certain Fleming"*: The Dutch Hans Lippershey applied for a patent for a telescope in 1608.

56. *"the Maker of the stars himself"*: In Van Helden 1989, 31.

58. *of Federico Cesi's Accademia dei Lincei*: Mario Biagioli gives a fascinating description of the social and cultural atmosphere associated with patronage, and Galileo's relation with Cesi and the Accademia dei Lincei. Biagioli 1993. I am grateful also to Stefano Gatti for helpful information on Cesi.

60. *that mountain to be more than four miles*: Given the limited quality of the data at the time, the estimate was not bad, but see also Adams 1932.

60. *When we compare Galileo's Drawings*: Galluzzi 2009 gives a detailed comparison of Galileo's drawings with modern observations.

60. *The Flight to Egypt*: The painting is in the Alte Pinakothek, Munich. For an interesting essay about the painting, see McCouat 2016. An online reproduction of the painting (which can be enlarged to see the lunar details) is at https://upload.Wikimedia.org/Wikipedia/commons/l/le/Adam_Elsheimer_-Die_Flucht_nacht_Ägypten%28AltePinakothek%29.jpg.

60. *by his friend Federico Cesi*: Elsheimer was a good friend of the German doctor and botanist Giovanni Feber, who in addition to being Cesi's friend, became a member of the Accademia dei Lincei in 1611.

60. *An intriguing story related*: The entire affair was described beautifully by Nicholas Schmidle in the *New Yorker*. Schmidle 2013.

61. *"If we examine the matter more closely"*: In *The Sidereal Messenger*, Van Helden 1989, 53.

62. *"What is so surprising about that?"*: Ibid., 55.

62. "she is movable and surpasses the Moon": Ibid, 57.

63. *"A great and wondrous sign appeared"*: Revelation 12:1.

64. *broke up into countless faint stars*: In Galileo's words in *Sidereal Messenger*: "the multitude of small ones is truly unfathomable."

66. *Being well aware of Bruno's tragic end*: Hilary Gatti published an interesting collection of essays on Bruno's richly diverse interests. Gatti 2011.

66. *recent estimates of the number of roughly*: For example, Petigura, Howard, and Marcy 2013.

67. *"Since they sometimes follow"*: *Sidereal Messenger*, Van Helden 1989, 84. William Shea 1998 and Noel Swerdlow 1998 give excellent summaries of Galileo's discoveries with the telescope and their implications for Copernicanism.

72. *A true explanation for the strange ears*: In the 1640s, Johannes Hevelius and Pierre Gassendi made many observations of Saturn. Hevelius and the famous architect Christopher Wren proposed incorrect models in 1656 and 1658, respectively. Huygens's suggestion of a flat ring appeared in his book *The Saturnian System*, which was published in 1659. Collected in Huygens 1888, vol. 15, 312. See also Van Helden 1974.

73. *between October and December 1610*: Gingerich 1984 and Peters 1984 discuss the phases of Venus.

74. *"Far from the least important"*: Clavius 1611–12.

76. *in the preface to his book* Bodies in Water: The translation of the book by Thomas Salusbury was edited by Stillman Drake. Galilei 1612.

78. History and Demonstrations Concerning Sunspots: An English translation appears in Drake 1957, 89.

79. *Galileo moved on to dismantling*: Bernard Dame 1966 presents the entire affair surrounding the controversy over sunspots. Dame 1966. See also Van Helden 1996. Translated excerpts from the three letters to Welser, with an introduction, appear in Drake 1957, 59.

81. *"The eternal mystery of the world"*: Einstein 1936.

CHAPTER 5: EVERY ACTION HAS A REACTION

83. *Galileo claimed that his new device—the telescope*: Spectacular images are in Galluzzi 2009.

84. *"Let us conclude, therefore, that he"*: Coresio 1612, cited in Shea 1972.

84. *"Before we consider Galileo's demonstrations"*: In di Grazia's *Considerazioni* (1612), which was reprinted in A. Favaro's *Opere di Galileo Galilei*, vol. 4, p. 385.

85. *"As if geometry in our day"*: From *Discourse on Bodies in Water*, cited in Shea 1972.

85. *wrote to Christopher Clavius*: Welser wrote from Augsburg, Bavaria, on March 12, 1610, one day before the *Sidereus Nuncius* was published. In Galluzzi 2017, 5.

86. *Giovanni Bartoli, wrote on March 27*: Cited in Heilbron 2010, 161.

87. *"His hair hung down"*: A letter of Horky to Kepler on April 27, 1610. In *Opere di Galileo Galilei*, vol. 10, pp. 342–43.

90. *"The conclusion is quite clear. Our Moon exists"*: Kepler wrote this in his *Dissertatio cum Nuncio Sidereo* of 1610.

92. *"Having presented to Her Majesty"*: Cited in Bucciantini, Camerota, and Giudice 2015, 168.

93. *"But you, O Galileo"*: Ibid., 190.

93. Assumption of the Virgin: Finer details of the lunar surface in this fresco can be seen online at: Flickr, accessed July 16, 2019, www.flickr.com/photos/profzucker /22897677200

93. *not as a smooth*: Booth and Van Helden 2000 give a detailed discussion of the depiction of the Moon in paintings.

94. *"It is true that with the spyglass"*: *Opere di Galileo Galilei*, vol. ii, pp. 92–93. Cited also in Van Helden 1989, iii.

96. *Ironically, some climate change deniers*: For a discussion of the fallacy involved in the Galileo gambit, see, for example, Mann 2016.

96. *Some 400 books*: Of 405 books, 234 were Italian, 56 French, 43 German, 22 English, and 50 from other countries. Of these, 160 were favorable to Galileo, 114 unfavorable, and 131 generally neutral. Drake 1967.

97. *The name was suggested by theologian*: Rosen 1947, 31.

97. *now known as the "Copernican principle"*: For an explanation, see, for example, Livio 2018.

100. *about 120 letters*: All described and analyzed beautifully by Sobel 1999, and translated and edited in Sobel 2001.

102. *he wrote his long and detailed* Letter to Castelli: Maurice Finocchiaro's translation of the slightly revised letter appears in Finocchiaro 1989, 49–54, and in Finocchiaro 2008, 103–9. In the *Opere di Galileo Galilei*, it is in vol. 5, pp. 281–88. In 2018 the original version was discovered. See description in chap. 6.

104. *"proved himself more perspicacious"*: Cited, for example, in Frova and Marenzana 2006, 475.

104. *"there is more ado to interpret"*: Appears in the essay entitled "Of Experience," online at: www.gutenberg.org/files/3599/3599.txt.

CHAPTER 6: INTO A MINEFIELD

107. *Galileo took the first, insightful step*: For a discussion of Galileo's achievements, see also Shea 1972; Brophy and Paolucci 1962.

108. *a model invented so as to "save the appearances"*: Many authors have discussed Galileo's interaction with the Church. In addition to sources mentioned already, here are a few more that I found very helpful: Blackwell 1991, Finocchiaro 2010, and McMullen 1998.

109. *Even as late as 1945*: See chap. 16 for more details on this episode.

110. *There has been a fascinating*: The story is beautifully described in Camerota, Giudice, and Ricciardo 2018.

112. *"I am most displeased that the ignorance"*: Castelli wrote this letter on December 31. For a more complete text, see, for example, Drake 1978, 239.

113. *the current climate change deniers*: It is important to clarify that atmospheric physicist Richard Lindzen, who is often described as a climate change denier, does not deny the reality of climate change. He is only not convinced about the role of humans in producing it, and about the actions that are proposed

to be taken to solve the problem. An overwhelming majority of the scientific community disagrees with Lindzen. For a very brief summary of the current thinking regarding climate change, see, for example: "Climate Change: Where We Are in Seven Charts and What You Can Do to Help," BBC News online, last modified April 18, 2019, www.bbc.com/news/science-environment-46384067. See also Schrag 2007. For Lindzen's minority opinion see, for instance, Lindzen, "Thoughts on the Public Discourse over Climate Change," *Merion West*, last modified April 25, 2017, https://merionwest.com/2017/04/25/richard-lindzen -thoughts-on-the-public-discourse-over-climate-change.

114. *"he [Barberini] would like greater caution"*: The letter from Ciampoli to Galileo was dated February 28, 1615. In *Opere di Galileo Galilei*, vol. 12, p. 146. A good description of the events is given in Shea and Artigas 2003 and in Fantoli 2012.

114. *Bellarmino suggested further that Galileo*: Described in detail in Blackwell 1991, 73.

117. *"Scripture serves us by speaking"*: Foscarini's publication appears as app. 6 in Blackwell 1991. The citation is from p. 232.

118. *"could not have appeared at a better time"*: *Opere di Galileo Galilei*, vol. 12, p. 150. Cesi added: "The writer counts all Linceans as Copernicans, though that is not so; all we claim in common is freedom to philosophize in physical matters."

119. *"it seems to me that Your Paternity"*: The translation appears in Finocchiaro 1989, 67.

119. *"Nor can one reply that this"*: Bellarmino's entire letter is reproduced in Fantoli 1996, 183–185, where it is also discussed, and also in Finocchiaro 1989, 67–69.

120. *Bellarmino's answer to Foscarini*: In addition to works mentioned already, there are interesting discussions about Bellarmino's opinions in Feldhay 1995, Coyne and Baldini 1985, Geymonat 1965, and Peia 1998.

123. *"It is replied that then everything"*: Galileo's unpublished notes from 1615, on Bellarmino's "Letter to Foscarini," appear as app. 9, section A, in Blackwell 1991.

CHAPTER 7: THIS PROPOSITION IS FOOLISH AND ABSURD

128. *"taking the former as the cause"*: Galileo's theory of the tides is discussed, for instance, in Wallace 1992 and in Shea 1998.

129. *"this proposition is foolish and absurd"*: In the consultor's report on Copernicanism from February 24, 1616. The report is online at "Galileo Trial: 1616 Documents," DouglasAllchin.net, accessed July 16, 2019, douglasallchin.net/galileo/library /1616docs.htm The phrase is cited in a number of Galileo biographies, including Reston 1994, 164.

130. *Events then followed in rapid succession*: Described in detail in Fantoli 1996 and Fantoli 2012.

130. *"At the palace of the usual residence"*: One translation of the description of this Injunction event (from February 26, 1616) can be found online at "Galileo Trial: 1616 Documents," DouglasAllchin.net, accessed July 16, 2019, douglasallchin.net /galileo/library/1616docs.htm. The original is in *Opere di Galileo Galilei*, vol. 19, pp. 321–22. The translation is in Finocchiaro 1989, 147–48, and Finocchiaro 2008, 175–76.

132. *the Congregation published its devastating decree*: The translation of the full text, from *Opere di Galileo Galilei*, vol. 19, pp. 322–23, appears in Finocchiaro 1989, 149, and in Fantoli 2012, 106.

133. *"especially in this century, for the present Pope"*: *Opere di Galileo Galilei*, vol. 12, 242. Translated in de Santillana 1955, 116.

134. *"We, Roberto Cardinal Bellarmino"*: This document was included with those of Galileo's trial in 1633, since that was when Galileo presented it. It appears in Pagano 1984, and the English translation is in Finocchiaro 1989, 153.

135. *Why did the Jesuit mathematicians?*: George Coyne 2010 also discusses this question.

CHAPTER 8: A BATTLE OF PSEUDONYMS

140. *"images and wandering simulacra"*: Galileo's argument (as it appears in Guiducci's "Discourse") is discussed in Drake and O'Malley 1960, 36–37. An excellent discussion on the comet controversy is in chap. 4 of Shea 1972.

141. *never advanced an actual theory*: David Eicher gives the modern view on comets. Eicher 2013.

142. *"I shall not pretend to ignore"*: Discussed in Galluzzi 2014, 251, and also in Drake and O'Malley 1960, 57.

146. *"What is this sudden fear"*: *Opere di Galileo Galilei*, vol. 6, p. 145. English translations are, for instance, in Langford 1971, 108, and in Fantoli 2012, 128.

147. *"Philosophy is to be studied, not for the sake"*: Russell 1912.

149. *"thanks to the subtle and solid"*: *Opere di Galileo Galilei*, vol. 6, p. 200. Translation in Fantoli 2012, 129.

150. *an ode, "Adulatio Perniciosa"*: Included, with its English translation, in Gattei 2019.

150. *"assuring you that you will find in me"*: Letter sent on June 24, 1623. In *Opere di Galileo Galilei*, vol. 13, p. 119.

151. *"a stupendous masterpiece of polemic"*: Geymonat 1965, 101.

CHAPTER 9: THE ASSAYER

153. *"In Sarsi I seem to discern the firm"*: In many ways, this text marked the beginning of modern physics. Princeton Institute for Advanced Study theoretical physicist Nima Arkani-Hamed said recently in an interview: "The ascension to the tenth level of intellectual heaven would be if we find the question to which the universe is the answer." Galileo started that quest; see Wolchover 2019. Extensive excerpts from *The Assayer* are in Drake 1957 and also Drake and O'Malley 1960.

155. *"If Sarsi wishes me to believe"*: *Opere di Galileo Galilei*, vol. 6, p. 340. Translation in Drake and O'Malley 1960.

156. *"To excite in us tastes, odors, and sounds"*: For a discussion of this topic from an epistemological perspective, see Potter 1993.

156. *to his subsequent troubles with the Church*: The speculation that atomism was the main reason for Galileo having been declared a heretic was developed in Redondi 1987. Most scholars don't accept this theory.

156. *"Let us be granted that my master"*: Opere di Galileo Galilei, vol. 6, pp. 116. Translation in Drake and O'Malley 1960, 71. The fact that Mars crosses the Sun's path was known to be a problem for the Ptolemaic model.

158. *The latter reportedly rushed to the Sun bookshop*: The entire event is described in Opere di Galileo Galilei, vol. 13, pp. 145, 147–48. See also Redondi 1987, 180.

158. *Italian scholar Pietro Redondi discovered*: The letter that he found is reproduced at the end of *Galileo Heretic* (Redondi 1987). Redondi's book describes the entire Galileo-Grassi conflict as part of a much larger social drama.

CHAPTER 10: *THE DIALOGO*

161. *"we cannot limit the divine power"*: According to the papal personal theologian, Cardinal Agostino Oreggi, Cardinal Barberini told this to Galileo. Oreggi 1629; cited in Fantoli 2012, 137.

162. *first by answering Francesco Ingoli's*: The English translation of Galileo's reply to Ingoli (from 1624) is in Finocchiaro 1989, 154–97, and discussed in Fantoli 1996, 323–28.

164. *"Here we have Galileo, who is a famous"*: Opere di Galileo Galilei, vol. 14, p. 103. Translated in Fantoli 1996, 336.

165. *this was not the end of the trials*: The ups and down of the process are described in detail in Fantoli 1996, Heilbron 2010, and Wootton 2010.

165. *"I agree to give the label"*: From Galileo's letter on May 3, 1631, to the Tuscan secretary of state. Translated in Finocchiaro 1989, 210–11.

166. Dialogue Concerning the Two Chief: There are several translations and commentaries on the *Dialogo*, such as Gould 2001, Finocchiaro 2014, and Finocchiaro 1997.

168. *"There were those who impudently asserted"*: Opere di Galileo Galilei, vol. 7, p. 29.

169. *"It is not from failing to take count"*: Opere di Galileo Galilei, vol. 7, p. 30; translated by Stillman Drake in Gould 2001, 6.

169. *Fashioned after Plato's dialogues*: Finocchiaro 1997 provides a selection from the *Dialogo*, with helpful commentary.

171. *To lambaste Galileo for both*: Koestler even wrote: "impostures like Galileo's are rare in the annals of science." Koestler 1989, p. 486.

171. *More recent, thorough analyses*: In particular, A. Mark Smith in 1985 and Paul Mueller in 2000 have shown that while the way Galileo presented his arguments was far from perfect (on both logical and completion grounds), once properly analyzed, the proof for the Earth's motion from sunspots was worth much more than the proof from the tides.

172. *"I know that if I asked whether God"*: Opere di Galileo Galilei, vol. 7, p. 488, translated by Stillman Drake in Gould 2001, 538.

173. *"I do not give these arguments the status"*: Opere di Galileo Galilei, vol. 7, p. 383. See also Gingerich 1986 for a concise summary of Galileo's contributions to astronomy.

CHAPTER 11: THE GATHERING STORM

176. *In early September Filippo Magalotti*: Fantoli 1996, chap. 6, describes in detail the sequence of events.

177. *"While we were discussing those delicate"*: Opere di Galileo Galilei, vol. 14, pp. 383–84. Months later, when Niccolini brought up the subject again, the Pope again exploded. See also Biagioli 1993, 336–37.

178. *"deceitfully silent about the command"*: Cited, for example, in Koestler 1990.

180. *He also wrote to a friend in Paris, Elia Diodati*: Diodati was born in Geneva but settled in France. He met Galileo during one of his trips to Italy, around 1620. Galileo wrote in 1636 that Diodati was his most cherished and true friend. After Galileo's death, Diodati kept in contact with Vincenzo Viviani.

181. *"May God forgive Signor Galilei"*: From a letter of the Tuscan ambassador Francesco Niccolini to the Tuscan secretary of state, Andrea Cioli. The letter was written on March 13, 1633. The translation appears in Finocchiaro 1989, 247.

CHAPTER 12: THE TRIAL

183. *After a few preliminary questions*: Among the many descriptions of the trial and its aftermath, I find those in Blackwell 2006, Finocchiaro 2005, Fantoli 2012, and de Santillana 1955 particularly illuminating.

184. *These discrepancies have spawned*: One of the reasons for the suspicions has been the fact that the document describing Seghizzi's intervention did not have the signatures of Galileo, Seghizzi, or any witnesses. Another was the fact that this document had conveniently been discovered just prior to the trial. The calligraphic analysis was performed by Isabella Truci of the National Central Library of Florence. Since the document presented only a summary, no signatures other than that of the notary were required.

185. *"I do not recall that such injunction"*: Opere di Galileo Galilei, vol. 19, p. 340. Translated in Finocchiaro 1989, 260, as part of the session on April 12, 1633.

189. *Melchior Inchofer, who was a strong opponent*: He also published a treatise entitled *A Summary Treatise Concerning the Motion or Rest of the Earth and the Sun, in which it is briefly shown what is, and what is not, to be held as certain according to the teachings of the Sacred Scriptures and the Holy Fathers.* Inchofer 1633.

189. *"Last night, Galileo was afflicted"*: The Italian version of the letter is in Beretta 2001. The translation here is from Blackwell 2006, 14.

190. *One Galileo scholar therefore suggested*: In particular, this thesis has been advanced in Blackwell 2006. Others, such as Heilbron 2010, were not convinced. In a private conversation, Michele Camerota told me that he believed there was an agreement between Maculano and Galileo. In a private conversation, Paolo Galluzzi suggested that this agreement is what may have led to the fact that Galileo served his prison sentence only under house arrest.

190. *"to deal extrajudicially with Galileo"*: Blackwell 2006, 224, argued that without the

plea bargain, it is difficult to understand why Galileo confessed in the second
session of the trial, given that his position after the first session was quite solid.
Fantoli 1996, 426, agreed.

191. *a clear demonstration of what intimidation*: As philosopher Albert Camus noted,
even Galileo, "who held a scientific truth of great importance, abjured it . . . as
soon as it endangered his life." Camus 1955, 3.

192. *"not introduced through the cunning"*: In addition to this confession, and after asking his judges to consider his poor health and advanced age, Galileo also asked
them to consider his honor and reputation against the slanders of those who
hated him. Finocchiaro 1989, 280–81.

CHAPTER 13: I ABJURE, CURSE, AND DETEST

193. *even some downright false*: Mostly from Lorini's and Caccini's claims. The accusation in Blackwell 2006 that the *Letter to Benedetto Castelli* was also falsified is not
true, as shown by the text in the original *Letter to Benedetto Castelli*, which was
discovered in 2018 (as described in chap. 6).

194. *The summary itself was most probably written*: This was at least the opinion of
Giorgio de Santillana. Santillana 1955, 284.

195. *In the first English biography of Galileo*: All copies but one were destroyed by the
Fire of London in 1666. Even that one copy was lost in the mid-nineteenth century, only to resurface temporarily in an auction between 2004 and 2007. Wilding
2008 gives a superb description of the history of the manuscript.

195. *about being caricatured as Simplicio*: Finocchiaro argues convincingly that this was
not the main reason for the trial. Finocchiaro 2005, 79.

196. *Add to this, that he [the Pope]*: Cited in Wilding 2008, 259.

196. *"decreed that the said Galileo"*: Cited in Langford 1966, 150, and also in Blackwell
2006, 22.

197. *"vehemently suspected of heresy"*: After formal heresy, this was the next crime in
terms of severity.

197. *"We are willing to absolve you"*: Translated in Finocchiaro 1989, 291.

197. *We do not know if Cardinal Francesco Barberini's*: A number of Galileo scholars believe that Francesco Barberini's absence (he was absent also on June 16) signaled
disapproval (for example, de Santillana 1955, 310–11). The other absentees were
Cardinal Caspar Borgia and Cardinal Laudivio Zacchia.

198. *"I abjure, curse, and detest"*: *Opere di Galileo Galilei*, vol. 19, pp. 402–6; translated in
Finocchiaro 1989, 292.

199. *A thorough investigation I conducted*: This "detective story" will be described elsewhere.

199. *in* The Italian Library: Baretti 1757.

CHAPTER 14: ONE OLD MAN, TWO NEW SCIENCES

202. *"As much as the news of Your Honor's"*: Letter of Maria Celeste on July 2, 1633. Slightly different translations can be found on the website *The Galileo Project*, as well as in Heilbron 2010, 327, and in Sobel 1999, 279. All the letters are in Sobel 2001.

203. *"What an unhappy place we live in"*: Letter of Pieroni to Galileo on August 18, 1636, cited in Heilbron 2010, 331.

203. *marked the final chapter of Galileo's scientific*: For those who can read Italian, one of the best books on Galileo and the scientific culture of his time is Camerota 2004.

206. *"Along a horizontal plane the motion"*: Galilei 1914, 215.

206. *"The Lord God is subtle"*: Einstein said it to Princeton mathematician Oswald Veblen in May 1921. It is now inscribed in the faculty lounge, 202 Jones Hall.

206. *"My purpose is to set forth"*: In Galilei 1914, "Third Day."

CHAPTER 15: THE FINAL YEARS

207. *diagnosed his condition as*: Discussed in Zanatta et al. 2015 and in Thiene and Basso 2011.

207. *"Alas, my good sir, your dear friend"*: Letter to Diodati on January 2, 1638, cited in Fermi and Bernardini 1961, 109, and (with a slightly different translation) in Reston, 1994, 277.

208. *"And lest some should persuade ye"*: Milton 1644; text appears, for example, in Cochrane 1887, 74.

209. *"I inquired in Leiden and Amsterdam"*: Letter of Descartes to Mersenne in November 1633, cited in Gingras 2017.

210. *Modern medical researchers have speculated*: Zanatta et al. 2015, Thiene and Basso 2011.

210. *"At the age of seventy-seven years"*: *Opere di Galileo Galilei*, vol. 19, p. 623. Reproduced in Fantoli 2012, 218.

211. *Even though his remains lay*: Galluzzi 1998 gives an excellent description of the fate of Galileo's remains. As Galileo's remains were moved from his original grave to the tomb in the Basilica of Santa Croce, the thumb, index, a middle finger, and a tooth were detached from Galileo's body. Those are now on display in glass bell jars at the Galileo Museum in Florence (Figure 10 in the color insert). The fifth lumbar vertebra was also removed and it is now at the University of Padua. John Fahie 1929 compiled a list of various memorials to Galileo.

CHAPTER 16: THE SAGA OF PIO PASCHINI

213. *than the tale of Monsignor Pio Paschini*: The story is told in Fantoli 2012, 228–32, in Blackwell 1998, 361–65, and in detail in Finocchiaro 2005, 275–77, 280–284, 318–37, and Simoncelli 1992.

214. Life and Works of Galileo Galilei: Paschini 1964.

214. *and producing a manuscript on January 23, 1945*: In Simoncelli 1992, 59.

214. *and was rejected as "unsuitable"*: Paschini never received any written report on the objections. In a letter he sent on May 12, 1946, to Giovanbattista Montini, deputy secretary of state at the Vatican, he complained, "I was extremely surprised and disgusted that I should have been accused of having produced nothing but an apology of Galileo. In fact, this accusation profoundly attacks my scientific integrity as a scholar and teacher."

214. *especially that with his friend Giuseppe Vale*: On May 15, 1946, Paschini wrote to Vale about the decision of the Holy Office: "It said that my work was an apology for Galileo; it made some comments on a few of my sentences; it objected that Galileo had not given the proofs of his system (the usual sophism); and it concluded that publication was not appropriate." Finocchiaro 2005, 323.

215. *The general impression one got from Lamalle's*: Lamalle 1964.

215. *"One can, therefore, legitimately regret"*: The Second Vatican Council approved this text on December 7, 1965, in the *Gaudium et Spes* (Latin for Joy and Hope), one of the four constitutions resulting from the Second Vatican Council.

216. *when Bertolla scrutinized the individual changes*: Bertolla 1980, 172–208.

216. *"to be directed against the Copernican"*: Cited in Finocchiaro 2005, 334.

216. *Citing an article from 1906*: Delannoy 1906, 358.

218. *"Deep Harmony Which Unites the Truths"*: John Paul II 1979.

218. *"World Takes Turn in Favor of Galileo"*: Koven 1980.

219. *"Paradoxically, Galileo, a sincere believer"*: John Paul II 1992.

219. *The* New York Times *announced*: Cowell 1992.

219. *The* Los Angeles Times *had a similar*: Montalbano 1992.

219. *"The fact that the Pope continues"*: Beltrán Marí 1994, 73.

CHAPTER 17: GALILEO'S AND EINSTEIN'S THOUGHTS ON SCIENCE AND RELIGION

221. *Letter to the Grand Duchess Christina*: The text and commentary can be found in Drake 1957, 145–216.

222. *a 2017 Gallup survey*: The support for creationism was the lowest in thirty-five years. Swift 2017. "Intelligent Designers" want to teach Darwinian evolution and creationist views simply as rival hypotheses. Gopnik 2013 presents an engaging discussion in the context of Galileo's biography.

222. *the controversy over which remains*: See Larson 2006, 1985.

222. *"to the field of religious doctrine"*: John Paul II, 1992.

223. *the age of the universe is now known*: Primarily from observations of the cosmic microwave background. Planck Collaboration 2016.

223. *"The concept of biological evolution"*: National Academy of Sciences president Bruce Alberts in the preface to *Science and Creationism: A View from the National Academy of Sciences*, 2nd ed., 1999.

223. *"The big bang, which nowadays"*: Pope Francis at the Plenary Session of Pontifical Academy of Sciences, Casina Pio IV.

223. *evolution by means of natural selection*: For a very clear exposition of the evidence for Darwinian evolution, see Coyne 2009.

225. *Climate change denial is fed mainly*: Yale researcher Dan Kahan studied what accounts for public opinion. See, for example: "What accounts for Public Conflict," www.culturalcognition.net/blog/2014/11/10/what-accounts-for-public-conflict-over-science-religiosity-0.html.

225. Emissions Gap Report (Nairobi: United Nations Environment Programme, November 2018), www.unenvironment.org/resources/emissions-gap-report-2018. The opinion of the vast majority of the scientific community on climate change is presented, for instance, by Schrag and Alley 2004, and Schrag 2007.

226. *in this context, nothing short of shocking*: For example, David Wallace-Wells 2019 paints a frightening picture of the potential impacts of climate change. Otto 2016 discusses the attacks on science. Crease 2019 analyzes how to address antiscience rhetoric.

226. *"By academic freedom, I understand"*: Made on March 13, 1954, Einstein Archives 28–1025.

226. *"Freedom of teaching and of opinion"*: The address was prepared for a meeting of university professors that never happened. Was published in Einstein 1950, 183–84.

226. *Einstein's opinions were more complex*: Max Jammer gives an excellent description, Jammer 1999.

227. *"The most beautiful experience we can have"*: The text is online at https://history.air.org/exhibits/einstein/essay.htm. Appeared in Einstein 1930.

228. *"I believe in Spinoza's God"*: Rabbi Goldstein commented that Einstein's reply "very clearly disproves . . . the charge of atheism made against Einstein." "Einstein Believes in 'Spinoza's God': Scientist Defines His Faith in Reply to Cablegram from Rabbi Here," *New York Times*, April 25, 1929, 60.

228. *which he wrote for the* New York Times Magazine: Einstein 1930.

229. *A disabled World War I veteran*: Letter to Einstein on September 11, 1940, Einstein Archive, reel 40–247.

230. *he told an anti-Nazi German*: Diplomat and author Hubertus zu Löwenstein. In Löwenstein 1968, 156.

230. *a letter that Einstein wrote*: The story associated with the letter is described in Livio 2018 and "The Word God Is for Me Nothing but the Expression and Product of Human Weakness," Christie's online, last modified December 12, 2018, www.christies.com/features/Albert-Einstein-God-Letter-9457-3.aspx.

230. *the judgment of Pope John Paul II*: See also John Paul II 1987.

230. *"Although I am not for religion"*: Cited, for example, in Miller 1997.

231. *We should allow for the coexistence*: Similar ideas have been expressed by Italian philosopher Dario Antiseri. See Antiseri 2005. A very interesting discussion of atheism is by Gray 2018. Jerry Coyne 2015 argued convincingly that attempts to *reconcile* the scientific and religious arguments (rather than allowing them to coexist in their parallel realms) are doomed to fail, because faith does not represent facts. On the other hand, Hardin, Numbers, and Binzley 2018 attempt to refute the concept that there is warfare between science and religion.

231. *and disallow only intolerance*: The new version of the International and Religious Freedom Act reads: "The freedom of thought, conscience, and religion is understood to protect theistic and non-theistic beliefs and the right not to profess or practice any religion."

CHAPTER 18: ONE CULTURE

233. *what author John Brockman dubbed*: Brockman 1995. C. P. Snow himself introduced the term "Third Culture" in the 1960s, but he referred to the social scientists.

233. *life expectancy in England*: Figures from the Office for National Statistics for 2015–17.

234. *succeeded in detecting gravitational waves*: The direct detection was made on September 14, 2015, by the LIGO and Virgo collaborations. Abbott et al., 2016.

235. *the humanities and the sciences are integral parts*: This topic is extensively analyzed and discussed in Pinker 2018. A great read. In a series of books edited by John Brockman (for example, Brockman 2015, 2018, 2019), Brockman compiled ideas from thinkers in a wide range of disciplines on particular concepts, thus effectively demonstrating the concept of one culture.

236. *Why does the universe exist?* Beautifully discussed in Holt 2013, in conversation with thinkers.

236. *science has already provided at least partial answers*: A detailed popular description of the history of the science in this topic is in Krauss 2017.

236. *we now know that our universe started*: Rees 1997, 2000 provides clear, accessible explanations of the cosmological parameters that determine the history and fate of our universe. Carroll 2016 gives a vivid description of humanity's place in the cosmos. Randall 2015 illustrates the intriguing connections that can exist between the universe's make-up and life on Earth.

236. *"no less foolish than that of a certain"*: Galileo wrote this as part of his response to delle Colombe and di Grazia in 1611. *Opere di Galileo Galilei*, vol. 4, p. 30–51.

237. *"Education needs to impart skills"*: In Nussbaum's excellent book *Not for Profit*. Nussbaum 2010.

237. *In so many works of art*: Tognoni 2013 describes many of these in detail.

Bibliography

Abbott, B. P., et al. (LIGO Scientific Collaboration and Virgo Collaboration). 2016. "Observation of Gravitational Waves from a Binary Black Hole Merger." *Physical Review Letters* 116:061102.

Adams, C. W. 1932. "A Note on Galileo's Determination of the Height of Lunar Mountains." *Isis* 17, 427.

Antiseri, D. 2005. "A Spy in the Service of the Most High." www.chiesa, accessed July 16, 2019, chiesa.espresso.repubblica.it/articolo/41533%26eng63oy .html?refresh;_ce.

Baretti, G. 1757. *The Italian Library. Containing an Account of the Lives and Works of the Most Valuable Authors of Italy.* London: A. Millar.

Bedini, S. A. 1967. "The Instruments of Galileo Galilei." In *Galileo Man of Science.* Edited by E. McMullin. New York: Basic Books.

Beltrán Marí, A. 1994. Introduction, in *Diálogo Sobre los Dos Máximos Sistemas del Mundo.* Madrid: Alianza Editorial.

Beretta, F. 2001. "Un nuove documento sul processo di Galileo Galilei: La Lettere di Vincenzo Maculano del 22 Aprile 1633 al Cardinale Francesco Barberini." *Nuncius* 16:629.

Bertolla, P. 1980. "Le Vicende del 'Galileo' di Paschini." In *Atti del Convegno di Studio su Pio Paschini nel Centenario della Nascita: 1878–1978.* Udine, It.: Poliglotta Vaticana.

Biagioli, M. 1993. *Galileo Courtier: The Practice of Science in the Culture of Absolutism.* Chicago: University of Chicago Press.

Bignami, G. F. 2000. *Against the Donning of the Gown: Enigma.* London: Moon Books.

Blackwell, R. J. 1991. *Galileo, Bellarmine, and the Bible*. Notre Dame, IN: University of Notre Dame Press.

―――. 1998. "Could There Be Another Galileo Case?" In *The Cambridge Companion to Galileo*. Edited by P. Machamer. Cambridge: Cambridge University Press.

―――. 2006. *Behind the Scenes at Galileo's Trial*. Notre Dame, IN: University of Notre Dame Press.

Booth, S. E., and Van Helden, A. 2000. "The Virgin and the Telescope: The Moons of Cigoli and Galileo," *Science in Context*, 13, 463.

Born, M. 1956. *Physics in My Generation*. Oxford: Pergamon Press.

Brockman, J. 1995. *The Third Culture*. New York: Simon & Schuster.

―――, ed., 2015. *What to Think About Machines That Think*. New York: Harper Perennial.

―――. ed. 2018. *This Idea Is Brilliant*. New York: Harper Perennial.

―――, ed., 2019. *The Last Unknown: Deep, Elegant, Profound Unanswered Questions About the Universe, the Mind, the Future of Civilization, and the Meaning of Life*. New York: William Morrow.

Brophy, J., and H. Paolucci. 1962. *The Achievement of Galileo*. New York: Twayne.

Bucciantini, M., and M. Camerota. 2005. "One More About Galileo and Astrology: A Neglected Testimony." *Glilaeana* 2:229.

Bucciantini, M., M. Camerota, and F. Giudice. 2015. *Galileo's Telescope: A European Story*. Translated by C. Holton. Torino, It.: Giulio Einaudi.

Camerota, M. 2004. *Galileo Galilei: E La Cultura Scientifica Nell'Età Della Controriforma*. Rome: Salerno.

Camerota, M., F. Giudice, and S. Ricciardo. 2018. "The Reappearance of Galileo's Original Letter to Benedetto Castelli." *Royal Society Journal of the History of Science*, last modified October 24. https://royalsocietypublishing.org/doi/10.1098/rsnr.2018.0053.

Camerota, M., and M. Helbing. 2000. "Galileo and Pisan Aristotelianism: Galileo's 'De Motu Antiquira' and the Quaestiones de Motu Elementorum of the Pisan Professors." *Early Science and Medicine* 5:319.

Camus, A. 1955. *The Myth of Sisyphus*. New York: Alfred A. Knopf.

Carroll, S. 2016. *The Big Picture: On the Origins of Life, Meaning, and the Universe Itself*. New York: Dalton.

Clavelin, M. 1974. *The Natural Philosophy of Galileo: Essay on the Origins and Formation of Classical Mechanics.* Translated by A. J. Pomerans. Cambridge, MA: MIT Press.

Clavius, C. 1611–12. *Opera Mathematica.* Vol 3., 75. In *Between Copernicus and Galileo, Christoph Clavius and the Collapse of Ptolemaic Cosmology.* Translated by J. M. Lattis 1994. Chicago: University of Chicago Press, 198.

Cochrane, R. 1887. *A Comprehensive Selection from the Works of the Great Essayists, from Lord Bacon to John Ruskin.* Edinburgh: W. P. Nimmo, Hay and Mitchell.

Cooper, L. 1935. *Aristotle, Galileo and the Tower of Pisa.* Ithaca, NY: Cornell University Press.

Coresio, G. 1612. *Operetta Intorno al Galleggiare de' Corpi Solidi.* Reprinted in Favaro, A. 1968. *Le Opere di Galileo Galilei*, Edizione Nazionale. Florence, It.: Barbera.

Cowell, A. 1992. "After 350 Years, Vatican Says Galileo Was Right: It Moves." *New York Times*, October 31.

Coyne, G. 2010. "Jesuits and Galileo: Tradition and Adventure of Discovery." *Scienzainrete.* Last modified February 2. www.scienceonthenet.eu/content/article /george-v-coyne-sj/jesuits-and-galileo-tradition-and-adventure-discovery /february.

Coyne, G. V., and V. Baldini. 1985. "The Young Bellarmine's Thoughts on World Systems." In *The Galileo Affair: A Meeting of Faith and Science.* Edited by G. V. Coyne, M. Heller, and J. Życiński. Vatican City State: Specola Vaticana. 103.

Coyne, J. A. 2009. *Why Evolution Is True.* New York: Viking.

———. 2015. *Faith Vs. Fact: Why Science and Religion Are Incompatible.* New York: Penguin.

Crease, R. P. 2019. *The Workshop and the World: What Ten Thinkers Can Teach Us About Science and Authority.* New York: W. W. Norton.

Csikszentmihalyi, M. 1996. *Creativity: Flow and the Psychology of Discovery and Invention.* New York: HarperCollins.

Dame, B. 1966. "Galilée et les Taches Solaires (1610–1613)." *Revue d'Histoire des Sciences* 19, no. 4; 307.

Delannoy, P. 1906. Review of Vacandard, *Études, Revue d'Histoire Ecclésiastique*, 7:354–61.

Drake, S. 1957. "Excerpts from *The Assayer*." In *Discoveries and Opinions of Galileo*. Translated and with an introduction and notes by Stillman Drake. New York: Anchor Books.

———. 1967. "Galileo in English Literature of the Seventeenth Century." In *Galileo Man of Science*. Edited by E. McMullin. New York: Basic Books, 415.

———. 1973. "Galileo's Experimental Confirmation of Horizontal Inertia: Unpublished Manuscripts (Galileo Gleanings XXII)." *Isis* 64: 290.

———. 1978. *Galileo at Work: His Scientific Biography*. Chicago: University of Chicago Press.

Drake, S., and C. D. O'Malley eds. and trans. 1960. *The Controversy on the Comets of 1618: Galileo Galilei, Horatio Grassi, Mario Guiducci, Johann Kepler*. Philadelphia: University of Pennsylvania Press.

Eddington, A. S. 1939. *The Philosophy of Physical Science*. New York: Macmillan.

Eicher, D. 2013. *Comets! Visitors from Deep Space*. Cambridge: Cambridge University Press.

Einstein, A. 1930a. "What I Believe: Living Philosophies XIII." *Forum* 84: 193.

———. 1930b. "Religion and Science." *New York Times*, November 9, 1930.

———. 1934. "Geometrie und Erfahrung." In *Mein Weltbild*. Frankfurt am Main, Ger. Ullstein Materialien.

———. 1936. "Physics and Reality." *Journal of the Franklin Institute* 221, no. 3 (March): 349–82.

———. 1950. *Out of My Later Years*. New York: Wisdom Library of the Philosophical Library.

———. 1953. Foreword in *Dialogue Concerning the Two Chief World Systems, Ptolemaic and Copernican*. Edited by S. J. Gould. Translated by S. Drake. Berkeley: University of California Press.

———. 1954. In *On the Method of Theoretical Physics, Ideas and Opinions*. Edited and transcribed by S. Bargmann. London: Alvin Redman.

Eisenstein, E. L. 1983. *The Printing Revolution in Early Modern Europe*. Cambridge: Cambridge University Press.

Ericsson, A. and, R. Pool. 2016. *Peak: Secrets from the New Science of Expertise*. New York: Houghton Mifflin Harcourt.

Fabris, D. 2011. "Galileo and Music: A Family Affair." In *The Inspiration of Astronomical Phenomena* 6. Edited by E. M. Corsini. *Astronomical Society of the Pacific Conference Series*, 441: 57.

Fahie, J. J. 1929. *Memorials of Galileo Galilei, 1564–1642: Portraits and Paintings Medals and Medallions Busts and Statues Monuments and Mural Inscriptions*. London: Courier Press.

Fantoli, A. 1996. *Galileo: For Copernicanism and for the Church*. 2nd ed. Vatican City State: Vatican Observatory Publications.

———. 2012. *The Case of Galileo: A Closed Case?* Translated by G. V. Coyne. Notre Dame, In: University of Notre Dame Press.

Favaro, A. 1929. *Le Opere di Galileo Galilei*, Ristampa Della Edizione Nazionale (Firenze, It.: G. Barbera).

Feldhay, R. 1995. *Galileo and the Church: Political Inquisition or Critical Dialogue?* Cambridge: Cambridge University Press.

Fermi, L., and G. Bernardini. 1961. *Galileo and the Scientific Revolution*. New York: Basic Books.

Finocchiaro, M. A. 1989. *The Galileo Affair: A Documentary History*. Berkeley: University of California Press.

———. 1997. *Galileo on the World Systems: A New Abridged Translation and Guide*. Berkeley: University of California Press.

———. 2005. *Retrying Galileo: 1633–1992*. Berkeley: University of California Press.

———. 2008. *The Essential Galileo*. Indianapolis: Hackett.

———. 2010. *Defending Copernicus and Galileo: Critical Reasoning in the Two Affairs*. Dordrecht, Neth.: Springer.

———. 2014. *The Routledge Guidebook to Galileo's Dialogue*. London: Routledge.

Frova, A., and M. Marenzana. 2000. *Thus Spoke Galileo: The Great Scientist's Ideas and Their Relevance to the Present Day*. Oxford: Oxford University Press.

Galilei, G. 1590. *De Motu Antiquiora, Le Opere di Galileo Galilei*. Vol. 1. Translated by I. Drabkin and S. Drake 1960. In *On Motion and On Mechanics*. Madison: University of Wisconsin Press.

———. 1612. *Discourse on Bodies in Water*. Translated by T. Salusbury. Edited by S. Drake 1960. Urbana: University of Illinois Press.

———. 1638. *Discorsi e Dimonstrazioni Matematiche intorno a Due Nuove Scienze Attenenti alla Mecanica & i Movimenti Locali.* In *Opere di Galileo.* Vol. 8. Translated by S. Drake 1974. *Two New Sciences.* Madison: University of Wisconsin Press.

———. (1638) 1914. *Dialogues Concerning Two New Sciences.* First published 1638. Translated in 1914 by H. Crew and A. de Salvio. New York: Macmillan.

———. (1610) 1989. *Sidereus Nuncius, or The Sidereal Messenger.* Translated and with commentary by A. Van Helden. Chicago: University of Chicago Press.

Galilei, V. (1581) 2003. *Dialogue on Ancient and Modern Music.* Translated by C. V. Palisca. New Haven, CT: Yale University Press.

Galluzzi, P. 1998. "The Sepulchers of Galileo: The 'Living' Remains of a Hero of Science." In *The Cambridge Companion to Galileo.* Edited by P. Machamer. Cambridge: Cambridge University Press, 417.

———. 2009. Editor. *Galileo: Images of the Universe from Antiquity to the Telescope.* Florence, It.: Giunti.

———. 2017. *The Lynx and the Telescope: The Parallel Worlds of Cesi and Galileo.* Translated by P. Mason. Leiden, Neth.: Brill.

Gattei, S. 2019. *On the Life of Galileo: Viviani's Historical Account and Other Early Biographies.* Princeton, N.J.: Princeton University Press.

Gatti, H. 2011. *Essays on Giordano Bruno.* Princeton, NJ: Princeton University Press.

Geymonat, L. 1965. *Galileo Galilei: A Biography and Inquiry into His Philosophy of Science.* Norwalk, CT: Easton Press.

Gingerich, O. 1984. "Phases of Venus in 1610." *Journal for the History of Astronomy* 15:209.

———. 1986. "Galileo's Astronomy." In *Reinterpreting Galileo.* Edited by W. A. Wallace. Washington, DC: Catholic University of America Press, 111–26.

Gingras, Y. 2017. *Science and Religion: An Impossible Dialogue.* Translated by P. Keating. Cambridge: Polity Press.

Gladwell, M. 2009. *Outliers: The Story of Success.* London: Penguin.

Gopnik, A. 2013. "Moon Man: What Galileo Saw," *The New Yorker*, February 3.

Gould, S. J., ed. 2001. *Galileo Galilei: Dialogue Concerning the Two Chief World Systems.* Translated by S. Drake. New York: Modern Library. First published 1953 by University of California Press (Berkeley, CA).

Gower, B. 1997. *Scientific Method: An Historical and Philosophical Introduction.* London: Routledge.

Gray, J. 2018. *Seven Types of Atheism.* New York: Farrar, Straus and Giroux.

Heilbron, J. L. 2010. *Galileo.* Oxford: Oxford University Press.

Holt, J. 2013. *Why Does the World Exist?: An Existential Detective Story.* New York: Liveright.

Huygens, C. 1888. *Oeuvres Complètes de Christiaan Huygens.* Le Haye, NL: Martinus Nijhoff.

Inchofer, M. 1633. *A Summary Treatise Concerning the Motion or Rest of the Earth and the Sun.* Rome: Ludovicus Grignanus. Translated in Blackwell 2005, 105–206.

Jammer, M. 1999. *Einstein and Religion: Physics and Theology.* Princeton, NJ: Princeton University Press.

John Paul II 1979. "Deep Harmony Which Unites the Truths of Science with the Truths of Faith." *L'Osservatore Romano,* November 26: 9–10.

———. 1987. "The Greatness of Galileo Is Known to All." in *Galileo Galilei: Toward a Resolution of 350 Years of Debate—1633–1983.* Edited by Cardinal P. Poupard. Pittsburgh: Duquesne University Press, 195.

———. 1992. "Faith Can Never Conflict with Reason." *L'Osservatore Romano,* November 4, 1–2.

Koestler, A. 1959. *The Sleepwalkers: A History of Man's Changing Vision of the Universe.* London: Arkana. First published 1959 by Hutchinson (London).

Koven, R. 1980. "World Takes Turn in Favor of Galileo." *Washington Post* online, October 24. www.washingtonpost.com/archive/politics/1980/10/24/world-takes-turn-in-favor-of-galileo/81b41321-9868-47f2-adfc-09f0a6477907/?utm_term=.256414b0f233.

Koyré, A. 1953. "An Experiment in Measurement." *Proceedings of the American Philosophical Society* 97:222.

———. 1978. *Galileo Studies.* Translated by J. Mepham. Atlantic Highlands, NJ: Humanities Press.

Krauss, L. M. 2017. *The Greatest Story Ever Told . . . So Far: Why Are We Here?* New York: Atria.

Lamalle, E. 1964. "Nota Introduttiva All' Opera." In Paschini 1964. Vol. 1, vii–xv.

Langford, J. J. 1966. *Galileo, Science and the Church.* Ann Arbor: University of Michigan Press.

Larson, E. J. 1985. *Trial and Error: The American Controversy over Creation and Evolution.* New York: Oxford University Press.

———. 2006. *Summer for the Gods: The Scopes Trial and America's Continuing Debate over Science and Religion.* New York: Basic Books.

Lennox, J. G. 1986. "Aristotle, Galileo, and 'Mixed Sciences.'" In *Reinterpreting Galileo.* Edited by W. A. Wallace. Washington, DC: Catholic University of America Press.

Livio, M. 2018. "Einstein's Famous 'God Letter' Is Up for Auction." *Observations* (blog). *Scientific American* online, last modified October 11. https://blogs.scientific american.com/observations/einsteins-famous-god-letter-is-up-for-auction.

———. "The Copernican Principle." In *This Idea Is Brilliant.* Edited by J. Brockman. New York: Harper Perennial, 185.

Löwenstein, Prinz H. Zu. 1968. *Towards the Further Shore.* London: Victor Gollancz.

Machamer, P. 1998. Introduction. In *The Cambridge Companion to Galileo.* Edited by P. Machamer. Cambridge: Cambridge University Press.

MacLachlan, J. 1973. "A Test of an 'Imaginary' Experiment of Galileo's." *Isis* 64:374.

Macnamara, B. N., D. Z. Hambrick, and F. L. Oswald. 2014. "Deliberate Practice and Performance in Music, Games, Sports, Education, and Professions: A Meta-Analysis." *Association for Psychological Science* 25:1608.

Mann, M.E. 2012a. "The *Wall Street Journal,* Climate Change Denial, and the Galileo Gambit." *EcoWatch,* last modified March 28. www.ecowatch.com/the-wall -street-journal-climate-change-denial-and-the-galileo-gambit-1882199616.html.

———. 2012b. *The Hockey Stick and the Climate Wars: Dispatches from the Front Lines.* New York: Columbia University Press.

McCouat, P. 2016. "Elsheimer's Flight into Egypt: How It Changed the Boundaries Between Art, Religion, and Science." *Journal of Art in Society.* Accessed July 17, 2019. www.artinsociety.com/elsheimerrsquos-flight-into-egypt-how-it-changed -the-boundaries-between-art-religion-and-science.html.

McMullin, E. 1998. "Galileo on Science and Scripture." In *The Cambridge Companion to Galileo.* Edited by P. Machamer. Cambridge: Cambridge University Press, 271.

McTighe, T. P. 1967. "Galileo's 'Platonism': A Reconstruction." In *Galileo Man of Science*. Edited by E. McMullin. New York: Basic Books.

Michelet, J. 1855. *Histoire de France: Renaissance et Réforme*. Paris: Chamerot.

Miller, D. 1997. "Sir Karl Raimund Popper." *Biographical Memoirs of Fellows of the Royal Society* 43:369.

Milton, J. 1644. "Areopagitica; A Speech of Mr. John Milton For the Liberty of Unlicenc'd Printing, To the Parliament of England." London.

Montalbano, W. D. 1992. "Earth Moves for Vatican in Galileo Case." *Los Angeles Times*, November 1.

Mueller, P. R. 2000. "An Unblemished Success: Galileo's Sunspot Argument in the Dialogue." *Journal for the History of Astronomy* 31: 279.

National Academy of Sciences 1999. *Science and Creationism: A View from the National Academy of Sciences*. Washington, DC: National Academics Press.

Nuland, S. B. 2000. *Leonardo da Vinci: A Life*. New York: Viking.

Nussbaum, M. 2010. *Not for Profit: Why Democracy Needs the Humanities*. Princeton, NJ: Princeton University Press.

Oreggi, A. 1629. *De Deo Uno Tractatus Primus*. Rome: *Typographia*, Rev. Camerae Apostolicae 194–95.

Otto, S. 2016. *The War on Science: Who's Waging It, Why It Matters, What Can We Do About It*. Minneapolis: Milkweed Editions.

Pagano, S. M., ed. 1984. *I Documenti del Processo di Galileo Galilei*. Vatican City: Pontifical Academy of Science.

Panofsky, E. 1954. *Galileo as a Critic of the Arts*. The Hague: Martinus Nijhoff.

Paschini, P. 1964. *Vita e Opere di Galileo Galilei*. Edited by E. Lamaelle. In *Miscellanea Galileiana*. Vatican City State: Pontifical Academy of Sciences.

Pera, M. 1998. "The God of Theologians and the God of Astronomers: An Apology of Bellarmine." In *The Cambridge Companion to Galileo*. Edited by P. Machamer. Cambridge: Cambridge University Press, 367.

Peters, W. T. 1984. "The Appearances of Venus and Mars in 1610." *Journal for the History of Astronomy* 15:211.

Peterson, M. A. 2011. *Galileo's Muse: Renaissance, Mathematics and the Arts*. Cambridge, MA: Harvard University Press.

Petigura, E. A., A. W. Howard, and G. W. Marcy. 2013. "Prevalance of Earth-size Planets Orbiting Sun-like Stars." *Proceedings of the National Academy of Sciences* 110:19273.

Piccolino, M., and N. J. Wade, 2014. *Galileo's Visions: Piercing the Spheres of the Heavens by Eye and Mind.* Oxford: Oxford University Press.

Pinker, S. 2018. *Enlightenment Now: The Case for Reason, Science, Humanism, and Progress.* New York: Viking.

Planck Collaboration 2016. "Planck 2015 Results: XIII. Cosmological Parameters." *Astronomy & Astrophysics* 594:A13.

Potter, V. G. 1993. *Readings in Epistemology: From Aquinas, Bacon, Galileo, Descartes, Locke, Berkeley, Hume, Kant.* New York: Fordham University Press.

Randall, L. 2015. *Dark Matter and the Dinosaurs: The Astounding Interconnectedness of the Universe.* New York: Ecco.

Redondi, P. 1987. *Galileo Heretic.* Translated by R. Rosenthal. Princeton, NJ: Princeton University Press.

Rees, M. 1997. *Before the Beginning: Our Universe and Others.* New York: Basic Books.

———. 2000. *Just Six Numbers: The Deep Forces That Shape the Universe.* New York: Basic Books.

Reeves, E. 2014. Evening News: Optics, Astronomy, and Journalism in Early Modern Europe. Philadelphia: University of Pennsylvania Press.

Reston, J., Jr. 1994. *Galileo: A Life.* New York: HarperCollins.

Romm, J. 2016. *Climate Change: What Everyone Needs to Know.* Oxford: Oxford University Press.

Rosen, E. 1947. *The Naming of the Telescope.* New York: Henry Schuman.

———. 1966. "Galileo and Kepler: Their First Two Contacts." *Isis* 57:262.

Russell, B. 1912. *The Problems of Philosophy.* London: Home University Library. Reprint, Oxford: Oxford University Press, 1997.

Russell, B. 2007. *A History of Western Philosophy.* New York: Simon & Schuster.

Santillana, G. de. 1955. *The Crime of Galileo.* Chicago: University of Chicago Press. Reprint, Chicago, IL: Midway, 1976.

Schmidle, N. 2013. "A Very Rare Book." *New Yorker Online*, December 16. www .newyorker.com/magazine/2013/12/16/a-very-rare-book.

Schmitt, C. B. 1969. "Experience and Experiment: A Comparison of Zabarella's View with Galileo's in *De Motu*." *Studies in the Renaissance* 16:80.

Schrag, D. P. 2007. "Confronting the Climate-Energy Challenge." *Elements* 3:171.

Schrag, D. P., and R. B. Alley. 2004. "Ancient Lessons for Our Future Climate." *Science* 306:821.

Segre, M. 1989. "Galileo, Viviani and the Tower of Pisa." *Studies in History and Philosophy of Science* 20, no. 4 (December): 435.

Settle, T. B. 1961. "An Experiment in the History of Science." *Science* 133:19.

———. 1983. "Galileo and Early Experimentation." In *Springs of Scientific Creativity: Essays on Founders of Modern Science*. Edited by R. Aris, H. T. David, and R. H. Stuewer. Minneapolis: University of Minnesota Press, 3.

Shea, W. 1998. "Galileo's Copernicanism: The Science and the Rhetoric." In *The Cambridge Companion to Galileo*. Edited by P. Machamer. Cambridge: Cambridge University Press, 211.

———. 1972. *Galileo's Intellectual Revolution: Middle Period, 1610–1632*. New York: Science History Publications.

Shea, W. R., and M. Artigas. 2003. *Galileo in Rome: The Rise and Fall of a Troublesome Genius*. Oxford: Oxford University Press.

Simoncelli, P. 1992. *Storia di Una Censura: "Vita di Galileo" e Concilio Vaticano II*. Milan, It.: Frando Angeli.

Smith, A. M. 1985. "Galileo's Proof for the Earth's Motion from the Movement of Sunspot." *Isis* 76:543.

Snow, C. P. 1959. *The Two Cultures*. Cambridge: Cambridge University Press. Reprint, Cambridge: Cambridge University Press, 1993.

Sobel, D. 1999. *Galileo's Daughter: A Historical Memoir of Science, Faith, and Love*. New York: Walker.

———, trans. and ed., 2001. *Letters to Father*. New York: Walker.

Swerdlow, N. M. 1998. "Galileo's Discoveries with the Telescope and Their Evidence for the Copernican Theory." In *The Cambridge Companion to Galileo*. Edited by P. Machamer. Cambridge: Cambridge University Press, 244.

—————. 2004. "Galileo's Horoscopes." *Journal for the History of Astronomy* 35:135.

Swift, A. 2017. "In U.S., Belief in Creationist View of Humans at New Low." Gallup online. Last modified May 22. https://news.lgallup.com/poll/210956/belief-creationist-view-humans-new-low.aspx.

Thiene, G., and C. Basso. 2011. "Galileo as a Patient." In *The Inspiration of Astronomical Phenomena 6.* Edited by E. M. Corsini. Astronomical Society of the Pacific Conference Series 441:73.

Tognoni, F., Editor, 2013. *Le Opere di Galileo Galilei,* Edizione Nazionale, Appendice Vol. 1, "Iconografia Galileiana." Florence, Italy: Giunti.

Van Helden, A. 1974. " 'Annulo Cingitur': The Solution of the Problem of Saturn." *Journal for the History of Astronomy* 5:155.

—————. 1996. "Galileo and Scheiner on Sunspots: A Case Study in the Visual Language of Astronomy." *Proceedings of the American Philosophical Society* 140:358.

Van Helden, A. and E. Burr. 1995. "The Galileo Project," online at galileo.rice.edu.

Vasari, G. 1550. *Lives of the Most Eminent Painters, Sculptors, and Architects.* A second expanded edition appeared in 1568. A very elegant modern edition of a few of the biographies is: *The Great Masters.* Translated by G. D. C. de Vere. Edited by M. Sorino. 1986, Hong Kong: Hugh Lauter Levin.

Viviani, V. 1717. *Racconto Istorico della vita di Galileo Galilei (Historical Account of the Life of Galileo).* First published in *Fasti Consolari dell' Accademia Fiorentina.* Edited by Salvino Salvini. Florence, Italy. (Included in Favaro's *Opere di Galileo Galilei.* Vol. 19, 597.) English translation in Gattei 2019.

Wallace, W. A. 1992. *Galileo's Logic of Discovery and Proof: The Background, Content, and Use of His Appropriated Treatises on Aristotle's Posterior Analytics.* Dordrecht, Netherlands: Springer.

—————. 1998. "Galileo's Pisan Studies in Science and Philosophy." In *The Cambridge Companion to Galileo.* Edited by P. Machamer. Cambridge: Cambridge University Press.

Wallace-Wells, D. 2019. *The Uninhabitable Earth: Life After Warming.* New York: Tim Duggan Books.

Weinberg, S. 2015. *To Explain the World: The Discovery of Modern Science.* New York: Harper.

Wigner, E. P. 1960. "The Unreasonable Effectiveness of Mathematics in the Natural Science: Richard Courant Lecture in Mathematical Sciences Delivered at New York University, May 11, 1950." In *Communications in Pure and Applied Mathematics*, 13, no. 1. Reprinted in Saatz, T. L., and F. J. Weyl, eds. 1969. *The Spirit and the Uses of the Mathematical Sciences.* New York: McGraw Hill.

Wilding, N. 2008. "The Return of Thomas Salusbury's *Life of Galileo*." *British Society for the History of Science* 41:241.

———. 2014. *Galileo's Idol: Gianfrancesco Sagredo and the Politics of Knowledge.* Chicago: University of Chicago Press.

Wisan, W. L. 1974. "The New Science of Motion: A Study of Galileo's *De Motu Locali*." *Archive for History of Exact Sciences* 13:103.

Wolchover, N. 2019. "A Different Kind of Theory of Everything." *New Yorker* online, February 19. www.newyorker.com/science/elements/a-different-kind-of -theory-of-every-thing?fbclid+IWAR0Kc47OS_NuxPaj40PKn9zt3N_VO_hBli- jrN114EDqTJT7ipyaHSMteCiyk.

Wolfflin, H. 1950. *Principles of Art History: The Problem of the Development of Style in Later Art.* Translated by M. D. Hottinger. New York: Dover.

Wootton, D. 1983. *Paolo Sarpi: Between Renaissance and Enlightenment.* Cambridge: Cambridge University Press.

———. 2010. *Galileo: Watcher of the Skies.* New Haven, CT: Yale University Press.

———. 2015. *The Invention of Science: A New History of the Scientific Revolution.* New York: Harper.

Zanatta, A., F. Zampieri, M. R. Bonati, G. Liessi, C. Barbieri, S. Bolton, C. Basso, and G. Thiene. 2015. "New Interpretation of Galileo's Arthritis and Blindness." *Advances in Anthropology* 5:39.

Photo Credits

Index

Page numbers in *italics* refer to illustrations.
Page numbers beginning with 241 refer to endnotes.

Elizabeth I, Queen of England, 10
Elsevier, Louis, 203
Elsheimer, Adam, 60
Emergency Civil Liberties Committee, US, 226
England, 94, 209, 224, 233–34
Eohippus (horse), 224
ephemerides, 77
Epicurus, 158
Ernest of Bavaria, Elector of Cologne, 89, 93–94
Essay Concerning Human Understanding (Locke), 156
Essays (Montaigne), 104–5
Euclid, 10, 20, 21
Europa (Jupiter satellite), 69
European Space Agency, Gaia space observatory of, 65
evolution, theory of, 108, 157, 222–24, 225, 254
  gradualism and, 29
  natural selection and, 15, 223–24

Fabricius, Johannes, 76
Fabrizio, Girolamo, 236–37
Fahie, John Joseph, 32
"fake news," 35
falling bodies, 53, 80, 84, 203–6
  laws of, 4, 11, 23–24, 44–46
  motion and, 23–24
  see also weights
Fantoni, Filippo, 29
Farnese, Cardinal Odoardo, 89
Favaro, Antonio, 139, 241
Febei, Pietro Paolo, 193, 194
First Amendment of US Constitution, 208
Flight to Egypt, The (Elsheimer), 60, 245
Florence, xiii, 13, 15, 17, 19, 26, 52, 54, 70, 87, 91, 92, 99, 101, 139, 162, 165–66, 175, 179, 208
Florentine Academy, 25, 26, 139
Florentine Camerata, 19
Foscarini, Paolo Antonio, 120–25, 136
  published letter of, 116–19, 132

France, 94, 163–64, 203, 209
  Loire Valley of, 92
Francis, Pope, 223, 254
Franciscan order, 209
François I, King of France, 42
freedom of religion, 217, 256
freedom of speech, 208–9
freedom of thought, 213, 217
friction, 206

Galen, 7, 20
Galilei, Cosimo Maria, 101
Galilei, Galileo, *see* Galileo
Galilei, Giulia Ammannati, 17, 19, 55
Galilei, Livia (daughter), 42, 44, 100
Galilei, Livia (sister), 17, 41, 242
Galilei, Michelangelo, 6, 17, 19, 41, 44, 89, 207, 242
Galilei, Vincenzo (father), 6, 17–19, 20, 21, 198, 210
  works of, 18, 242
Galilei, Vincenzo (son), 23, 44, 100–101, 180, 208, 210
Galilei, Virginia (daughter), *see* Maria Celeste, Sister
Galilei, Virginia (sister), 17, 41, 55, 242
Galileo:
  analytical skills of, 27
  art, poetry, and music as interests of, 6–7, 15
  artistic training of, 79, 233
  biographies and other works on, xiii, 2, 5–6, 19–20, 21, 201, 213–17, 237
  birth of, 5, 42, 214, 242
  blindness of, 207–8, 237
  censure and house arrest of, 12, 42, 82, 100, 136, 201–2, 207
  character and personality of, xiii, 3–4, 17, 19, 37, 43–44, 78–79, 88, 90, 140, 144, 168, 173
  children of, 12, 42, 44, 99–100, 192, 201–2, 207
  compartmentalization rejected by, 236–37